W9-BYY-987

Experimental and Surgical
Technique in the Rat

To our wives and parents

Experimental and Surgical Technique in the Rat

Second Edition

H. B. WAYNFORTH
SmithKline Beecham Pharmaceuticals, UK

P. A. FLECKNELL
Comparative Biology Centre, Medical School,
University of Newcastle upon Tyne, UK

ACADEMIC PRESS

Harcourt Brace Jovanovich, Publishers

LONDON · SAN DIEGO · NEW YORK
BOSTON · SYDNEY · TOKYO · TORONTO

ACADEMIC PRESS LIMITED
24/28 Oval Road
London NW1 7DX

United States Edition published by
ACADEMIC PRESS INC.
San Diego, CA 92101

This book is printed on acid-free paper

A catalogue for this book is available from the British Library
ISBN 0–12–738851–6

Typeset by Photo·graphics, Honiton, Devon
Printed in Great Britain by St Edmundsbury Press Limited, Bury St. Edmunds,
Suffolk

PREFACE TO FIRST EDITION

In 1969 I wrote a chapter on experimental animal techniques (Waynforth, 1969). During my research for material for the chapter and subsequently, it became abundantly clear that although the smaller laboratory animals had been used in scientific research for many decades, not one detailed and concise book on general techniques in the individual species seemed to be readily available. Although most of the techniques are known, these have been published in individual papers scattered throughout the scientific literature, often in obscure or small circulation journals, and therefore do not command great attention. It has also struck me as curious that, considering the widespread use of experimental animals, neither in the United Kingdom nor to my knowledge elsewhere in the scientific world is there a national course of formal training in small animal experimentation, and experimental and surgical techniques. This, together with the paucity of specific literature on these topics, seems a ludicrous situation, particularly where, as in the United Kingdom, young scientists with no previous knowledge of the use of laboratory animals have to be licensed to perform animal experiments and are, by implication, legally expected "overnight" to become diagnosticians, experimentalists, pharmacists, anaesthetists and surgeons! This book, therefore, is an attempt at a remedy, and I would feel my task complete if it were to go some way towards enlightening the young, and even the more mature scientist with little knowledge of animal techniques, in the general principles of experimental small animal usage. Although this book is concerned specifically with the rat, many of the techniques described are directly applicable, or would be so with only slight modification, to the mouse. Some of the techniques could also be adapted for the guinea-pig.

It became apparent very early on that I could not hope to cover all aspects of experimental technique in the rat, and that I would have to be selective for those techniques which I felt had the greatest application and which I could have the most confidence writing about. I have been fortunate in having personal experience in performing, at one time or another, the majority of the techniques described. In other cases I have tried to gain some idea of the difficulties by asking my colleagues to show me the techniques, or by practising on dead animals, techniques gleaned from the literature. I have purposely not considered certain areas such as the alimentary tract because these have been covered adequately elsewhere (see Lambert, 1965).

I am indebted to many people for the help they have given me over the years in showing me how to carry out various techniques. I also appreciate the help given, often unwittingly, by my scientific colleagues both from this Institute and elsewhere. They have often asked me perplexing questions about animal usage which have caused me to attempt to find an answer, often successfully, and therefore add to my own education. I am also grateful to these same colleagues for asking me to collaborate in several scientific projects where I have gained further expertise and understanding of the animal experimental techniques that I have been asked to perform. My gratitude also extends to the staff of my section and of the Animal Unit who have invariably made my tasks much easier. Inevitably certain people have played a particularly important role, and I would like to express my gratitude to Dr Hugh Jones for initiating and encouraging my interest in experimental animal technique during the early part of my career, and for the time he has devoted and the precise way in which he has corrected and criticised the manuscript and offered many helpful suggestions. The excellent photographs were taken by Mr Brian Rice of the Department of Medical Photography, to whom I extend my sincere thanks, and were processed in this department by kind permission of Mr Ray Phillips. The difficult task of illustrating the surgical techniques fell to Jennifer Middleton (Mrs J. Halstead) to whom I express my admiration and gratitude for the perspicacity of the drawings. Lastly, I have been fortunate in being able to keep the secretarial work in the family by calling upon the professional typing ability of my cousin Mrs Barbara Bell, to whom I am most grateful, and my wife Frances, who in addition gave me much encouragement in my task.

Bryan Waynforth
January 1979

PREFACE TO SECOND EDITION

The reasons for writing this book in 1980 were because of the lack of detailed published information on methods of experimentation on the rat and the lack of suitable training courses in animal experimentation. Since then, books incorporating a variety of methods have appeared but either the range has been limited or the details have been sparse. Consequently this book still serves a useful purpose and indeed its continued use by investigators has encouraged the publishers to suggest a second edition.

The intention of the first edition was to help investigators achieve the best experimental animal preparation for their work. From letters received from readers, it would seem that the book was successful in doing this in many cases. The intention for this second edition is of course identical but with the addition of a greater emphasis on animal welfare. Consequently, present-day concepts of more humane ways to carry out certain techniques have been incorporated. Co-authorship of this edition also reflects this approach and also allows the incorporation of the progress that has been made in both general and specialist knowledge since 1980. Two new chapters have been written, on principles of surgery, and anaesthesia and analgesia, new techniques have been added and other techniques have been refined or expanded.

A particularly pleasing development over recent years has been the widespread establishment of training courses for investigators and in some countries this has become mandatory. However, in many cases, these courses have not developed to their full potential. In some countries, such courses concentrate only on handling of animals and theoretical aspects such as those on the legislation which has been enacted to control animal experimentation. In other countries, such as the United States of America, live animals are used to practise experimental and surgical procedures whereas in the

United Kingdom this is prohibited except, curiously, for attaining skill in microsurgery! No doubt one day wide consensus will be reached on what is the right approach, but it is our belief that investigators should have training in experimental procedures, anaesthesia and surgery before being allowed to use animals for scientific purposes. We hope that this book will make a positive contribution to such training when courses become fully developed.

The additional techniques described in this edition are not only the result of our own knowledge but we are fortunate in having colleagues who have willingly and unselfishly contributed. The following have our sincere thanks: David Gask (tail and jugular vein injection and bleeding in conscious rats); Eric Karren (intra-articular and subplantar injections); Robin Buckingham (catheterisation of the femoral artery); Pieter Groot (catheterisation of the mesenteric lymph duct); Nick Bowring (administration of drugs by inhalation). We have had numerous discussions with, and help from, other colleagues and in particular Heather Elliott, Tim Morris, and Chris West. We acknowledge the encouragement given by Dr Howard (Bud) Hughes, Vice President of Laboratory Animal Science, SmithKline Beecham during the production of this edition. The additional photography and drawings which so clearly enhance this second edition were kindly supplied by the staff of the Audio-Visual Centre, University of Newcastle upon Tyne, Ashley Waddle and by Nigel Webb, Sally Bradbury, and Chris David of the Department of Graphics and Photography, SmithKline Beecham.

We are especially grateful to Mrs Margaret Canham for typing the manuscript and coping with the idiosyncrasies of the two authors, and to Mrs Anne Al-Jumaili for assisting with the preparation of the final text.

Bryan Waynforth

Paul Flecknell

May 1991

INTRODUCTION

The first recorded instance of the use of rats in scientific research was probably in 1856, in a paper by J. M. Philipeaux entitled "Note sur l'extirpation des capsules surrénales chez rats albinos (*Mus rattus*)" (*C.R. Acad. Sci., Paris* **43**, 904). Since then the rat has undoubtedly become the premier animal in experimental work. In the United Kingdom 78% of all animal experiments performed during 1990 were carried out on the rat and mouse, and nearly one million rats alone were used.

There is no mystery as to why the rat commands such a high standing in experimental work. It breeds easily, profusely and continuously during its normal reproductive lifespan in the laboratory, is easily handled with practice, and can be housed in large numbers in a relatively confined area. Moreover, being a small animal, it is both economic and practical to use large numbers in an experiment and thereby lend greater statistical validity to the results. Much is now known about its physiology, anatomy, genetics and behaviour, and meaningful results can be obtained in rats which, if interpreted with care, can be extrapolated to man.

Because of the ubiquitous use of the rat, it is incumbent upon the investigator to know how to use the animal in the best possible way to achieve the most significant result. Paradoxically most scientists carry out no more than a handful of technical manipulations on the rat during their entire scientific career, and this may give the misleading impression that anything more than a superficial acquaintance with the theoretical and practical aspects of technique is unnecessary. However, a glance at some of the factors that can influence results (see p. 5) shows that experimental success or failure depends on the correct usage of the experimental animal. Such knowledge starts with the more mundane, but no less important skills, such as the correct handling of animals. Proper and frequent handling causes rats to become docile and easier to work with, reducing the chance of the investigator being scratched or bitten and consequently becoming agitated and spoiling his/her work. As an investigator will invariably transmit his or her nervousness, all animals should be approached slowly and steadily with at least an outward show of confidence! To attempt to catch small laboratory animals from behind in a somewhat indecisive fashion is extremely ill-advised. Rats which are properly handled suffer less stress and are consequently both physiologically and biochemically more 'normal' than those that have been treated roughly. This is important as stress increases the variability of the results in an experiment.

A knowledge of experimental technique allows the investigator to choose the best practical approach in the animal and avoids time being wasted in considering an experiment for which either no suitable techniques are available or where the degree of difficulty in carrying out the technique does not justify its use in a particular situation.

The same is true regarding the need for a wide understanding of surgical technique in the rat. Surgery in the rat has developed many techniques, some simple, others sophisticated and intricate. This book serves as an introduction for the new researcher to the range of techniques currently available, and is written in a way that allows it to be used as a laboratory manual. To this end, particular attention is placed on the anatomy of the rat, details of the anaesthesia and surgery involved, and the postoperative care required. A good knowledge of anatomical relationships is essential as this avoids the accidental manipulation of extraneous organs and tissues which could result in unwanted complications such as haemorrhage or nerve damage. The precise surgical technique adopted is a matter of personal choice and the skill of the operator. In this respect modification of the surgical techniques described in this book is to be expected, and indeed welcomed, as an indication of the greater degree to which the investigator has become involved. It is important that the surgically prepared animal is not allowed to suffer postoperatively from stress derived from incompetent technique or an ignorance of surgical care, all too possible if the investigator is ill-concerned with the fundamentals on which proper small animal surgery is based. Expertise in this field brings its own rewards, such as the longer survival of animals and the production of an animal which is healthier and which responds more accurately to subsequent treatment.

The performance of a technique greatly benefits from practice, and the importance of this cannot be stressed too strongly. In the United

Kingdom, the Animals (Scientific Procedures) Act, 1986, prohibits the use of live animals for the sole purpose of attaining manual skill except in the case of microsurgery. It is necessary, therefore, to use dead animals, and a useful degree of dexterity and confidence will be gained by using freshly killed rats.

Finally, no introduction into the use of animals for experimental purposes would be complete without encouraging the reader to adopt an humane approach. In practical terms this means treating animals compassionately, having regard to their welfare and, in particular, reducing all unnecessary pain and suffering. Unfortunately this attitude calls for a personal and emotive commitment on the part of the scientist, and therefore the question of the humane treatment of laboratory animals may often be given a low priority. The scientist should remember that pain is stressful, and the results of any or all of his or her experiments will be of questionable value unless he or she takes the trouble to either minimise or remove this source of variation. Therefore, to be scientific requires the humane approach, and to be humane is to be scientific.

CONTENTS

Chapter Two. Methods of Obtaining Body Fluids

Chapter Three. Anaesthesia and Postoperative Care

Chapter Four. Surgical Technique

Chapter Five. Specific Surgical Operations

Chapter Six. Miscellaneous Techniques

Chapter Seven. Vital Statistics and Miscellaneous Information

Chapter One

ADMINISTRATION OF SUBSTANCES

1. GENERAL ASPECTS OF THE ADMINISTRATION OF DRUGS AND SUBSTANCES

A number of questions often arise concerning the suitability of solutions for injection into an animal: "what volume can be used?", "will the pH affect the animal adversely?", "can a suspension be administered intravenously?" and so forth. Unfortunately precise answers are often not forthcoming, principally because of variability between animals. However, the following generalisations can be made and used as a guide for the investigator in order to overcome most of the problems that arise.

1.1. Solvents for Injection

Many substances can be injected conveniently in distilled water or physiological saline (0.9% sodium chloride). Both are suitable vehicles physiologically, but saline, which is isotonic with body fluids, is preferable (subcutaneously, distilled water causes pain, and intravenously, produces some haemolysis). Other "physiological" solvents which can be used are phosphate buffered saline (PBS), balanced salt solution (e.g. Hanks

1

BSS) and various tissue culture media. For reasons of solubility or rate of absorption, some substances may require a more complex solvent to render them suitable for administration. One or more of the following materials combined with distilled water, or saline can be tried: 60% (v/v) propane-1:2-diol (propylene glycol), 0.5% (w/v) carboxymethyl cellulose, 10% (v/v) Tween 80 (polyoxyethylene (20) sorbitan mono-oleate), 10% (v/v) ethyl alcohol, and 50% (v/v) dimethylformamide or dimethylsulphoxide. All of these vehicles can be administered by any of the injection routes available, but the concentrations mentioned are the maximum practicable, and in many cases it is possible and indeed desirable that lower concentrations should be used. The subject has been fully discussed by Yalkowsky (1981).

To the question "can solvents such as ethanol, acetone, benzene, carbon tetrachloride, dimethylformamide etc. be used in undiluted form?", the answer is a qualified "yes". However, these solvents are extremely toxic in neat form and must be used in minute quantities. Also, it depends by what route they are required to be used. Neat acetone or benzene is often used to apply materials to the skin without apparent harm. Acetone can be injected into an organ, but even in minute quantities will cause some cell death in the immediate vicinity of the injection. Neat solvents should only be used as a last resort.

Vegetable oils (e.g. olive oil, arachis (peanut) oil) are suitable for injection when lipid-soluble substances are to be administered. Absorption is delayed when oil is used as the solvent. Oil cannot be injected intravenously. If it is necessary to inject lipoidal substances intravenously then it may be possible to do so in a 15% (v/v) oil–water emulsion using, for example, lecithin as the emulsifier (Schurr, 1969).

Materials can be injected as a suspension. However, dosage will not be precise because of the tendency for the suspended particles to sediment. The suspended particles should be finely divided if the intravenous route is to be used. Flocculent suspensions can be more evenly prepared if a drop or two of Tween 80 is added. It should be noted that if injected intravenously, the particles will be filtered out in the capillary beds of the extremities and the lung, modifying the distribution of the injected material and sometimes causing pulmonary distress to the animal.

1.2. Volumes for Injection

The maximum quantity of a solution that can be injected into a rat will depend mainly on the route of administration. A list of suggested volumes for injection is given in Table 1. In some instances, e.g. subcutaneous (s.c.) and intramuscular (i.m.) injections, a greater volume can be administered if it is divided between several sites. However, for i.m. injection, no more than 0.20 ml for the adult rat will be retained at any one site of injection. The volume of material that can be administered by intravenous (i.v.) injection and the rate at which it can be injected will vary considerably depending upon the pH of the solution, its osmotic strength and whether it is likely to have any specific physiological effects. Rapid intravenous injection of up to 1 ml of isotonic saline can be undertaken in a 200 g rat without causing clinically significant cardiovascular disturbances. If the injection is made over a longer period, then larger volumes can be administered safely, for example 2–3 ml of material can be infused over a 5 minute period. A

Table 1. Suggested maximum volumes for injection into an adult (200 g) rat. These volumes should be reduced if the material is likely to be irritant to tissues.

Route	Maximum volume	Comments
s.c.	Up to 5 ml	Large volumes will be absorbed relatively slowly
i.m.	0.2 ml	Use more than one site. Large volumes disrupt muscle fibres, may not be retained within the muscle, and may cause pain
i.p.	Up to 10 ml	Only material in isotonic fluid should be administered in large volumes. Much smaller volumes should be used for irritant material
i.v.	2 ml	Inject over 1–2 min
Intradermal	0.05 ml	
Tracheal	40 μl	General anaesthesia required
Intragastric	5 ml	

very rapid injection, even of the animal's own blood, can produce cardiovascular failure and can be lethal.

Injection by the intraperitoneal (i.p.), subcutaneous (s.c.) and intragastric (gavage) routes can be carried out quickly. However, during administration by gavage care should be taken that the fluid remains in the stomach and is not regurgitated into the mouth where it will be spat out, or into the trachea, which may be fatal. Intramuscular and intraorgan injections must be done slowly, though the small volumes employed will not require a lot of time. In general, the smallest volume should be used that is compatible with solubility of the compound and accuracy of the dose.

1.3. pH of Injection Solutions

Rats can tolerate the administration of materials within a fairly wide range of pH. For all routes of administration, a working range is in the region of pH 4.5–8.0. Solutions of greater acidity can be tolerated orally if they do not exceed the equivalent of 0.1 N HCl, but alkaline solutions are not accepted well by the stomach. The widest tolerance to pH is shown by the i.v. route because of the buffering capacity of the blood, followed by the i.m. and then the s.c. routes.

1.4. Absorption of Injected Substances

The rate of absorption influences the time-course of drug effect, and is an important factor in determining drug dosage. Absorption from all sites of administration is dependent on drug solubility. Drugs given in solution are absorbed more quickly than those given in solid form, e.g. a suspension, and also local conditions alter solubility. For instance, at the low pH in the stomach acidic drugs may be absorbed slowly because they precipitate in the gastric fluids. The concentration of a drug also influences its rate of absorption. Drugs injected in solutions of high concentration are absorbed more rapidly than those in low concentration. The area of the absorbing surface to which a drug is exposed is an important determinant of the rate of drug absorption. Absorption is virtually instantaneous after i.v. injection. The peritoneal cavity offers a large absorbing surface, and aqueous drugs injected intraperitoneally are absorbed rapidly but about four times slower than after an i.v. injection. Absorption from an i.m. site is fairly rapid, and quicker than after s.c. injection. Substances which would cause considerable irritation if given subcutaneously can be injected safely into muscle tissue. The rate of absorption after administration by gavage is variable, and dependent on such factors as whether the drug is acidic or basic, whether it is ionised or not, the solubility and concentration of the drug, and even the rate of gastric emptying. In general, absorption of drugs from the alimentary tract is slower than after a s.c. injection. Few drugs are readily absorbed after application to the skin; absorption by the epidermis is proportional to the lipid solubility of the drug. Absorption through the skin can be increased by suspending the drug in oil and rubbing this into the skin. Once the drug has penetrated the epidermis it meets little resistance from the dermis which is permeable to many solutes.

1.5. Factors that May Modify Dosage

The investigator should be aware that the response to a given chemical or biological agent may well be altered by environmental and biological factors (Pakes et al., 1984; Fox, 1986; Bhatt et al., 1986). Some of these factors are presented below, and a number of examples will be given to illustrate the changes that can occur and to emphasise to the investigator the importance of giving careful consideration to any drug administration schedule that he might be considering. The effect of such variable factors can only be determined empirically.

1.5.1. Temperature

Rats are more susceptible to sodium pentobarbitone at low temperatures. A difference of room temperature from 24°C to 20°C for 1 week causes a 10-fold change in the cardiotoxic response of rats to isoprenaline. Rats fasted for 24 hours after being kept at 32°C for 3 months and then placed in a maze may never find a given food reward, in contrast to rats kept at 12°C or 23°C.

1.5.2. Light

Photosensitising agents such as haematoporphyrin have a fatal concentration in sunlight which is safe in the dark. Absence of sunlight is essential in certain vitamin D studies. Variation in light–dark periods can cause changes in the reproductive cycle and response to some compounds such as anaesthetics will vary during different phases of the photoperiod.

1.5.3. Emotion

Sodium pentobarbitone-treated rats, in the presence of a rabbit, will stay awake longer. A change of cage coupled with isolation (e.g. rats placed in metabolism cages) inhibits normal feeding and defaecation for at least 24 hours. Noise may affect various physiological functions, e.g. the noise from a normal fire alarm can alter reproductive capacity.

1.5.4. Diet

A low protein diet decreases the toxicity of the cancer-inducing drug, N-dimethylnitrosamine. Rats fed a cachexigenic diet develop a relative resistance to sodium pentobarbitone injected intraperitoneally. Rats fed a diet *ad libitum* develop many more tumours in old age than those fed 80% of the *ad lib*. food intake.

1.5.5. State of the Drug

In general, alcoholic solutions are more stable than aqueous ones. Sesame oil by itself decreases prostate and seminal vesicle weight and may therefore mask the effect of added androgens. N-methylnitrosourea, a cancer-inducing chemical, decomposes rapidly at alkaline pH. The solvent dimethylsulphoxide increases the toxicity of many quaternary ammonium compounds.

1.5.6. Route of Administration and Rate of Injection

Certain barbiturates and also snake venoms are more effective after oral than after parenteral administration. Injection of diisopropylfluorophosphate by the femoral vein route causes an increase in the motility of the gastrointestinal tract and in defaecation; injection via the femoral artery results in fasciculations of the leg muscles whereas the lowest incidents of effects result from administration via the hepatic portal vein.

1.5.7. Strain, Age and Sex

Different rat strains show different toxicities and tumour incidences after treatment with N-dimethylnitrosamine. Male rats are more resistant to the effects of nicotine and to sodium pentobarbitone anaesthesia than are female rats. Alloxan diabetes is easier to produce in older rats.

1.5.8. Biological Rhythms

Rats respond to an i.v. injection of histamine with a decrease in blood pressure, followed by a marked rise. This pattern, even in rats kept in the controlled environment of an animal house, is seen in autumn but does not occur in the summer.

1.5.9. Population Density

The response of rats to drugs and other parameters (e.g. X-irradiation) is often markedly affected by whether the animals are housed singly or in various sized groups.

1.5.10. Disease

Nitrosoheptamethylamine induces a 17% incidence of lung tumours in germ-free rats but an 83% incidence in conventional animals. Sendai virus infection affects the respiratory and immune systems, among others, and therefore complicates investigations that focus on these systems. *Pseudomonas aeruginosa* infection reduces the life-span of rats subjected to various surgical and immunosuppressive procedures.

2. ROUTES AND METHODS OF ADMINISTRATION

2.1. Apparatus for Injections

2.1.1. Hypodermic Syringes and Needles

Syringes are made of either glass or plastic. Plastic disposable syringes are widely used and have the advantage that they are supplied individually packaged in a sterile condition. They cannot be used, however, with solvents such as acetone. Most syringes have a Luer tip but some older glass syringes may be encountered with a Record fitting, which requires needles with a different diameter hub. In addition, there is a Luer–Lok fitting which prevents a Luer needle from being forcibly ejected from the syringe.

For most laboratory work, syringes with manually operated plungers are generally used. However, for blood withdrawal special vacuum-containing syringes are available in which blood is withdrawn automatically (Fig. 1). Such syringes probably have a very limited use in an animal as small as the rat, but can be useful for cardiac puncture.

Syringes are made in a variety of sizes to accommodate different volumes. The more usual sizes have a volume range from 1 to 60 ml and are variously calibrated. The tuberculin syringe, which is a long thin 1 ml syringe calibrated in 10 μl divisions, is commonly used for delivery of small volumes, but a disposable 0.5 ml insulin syringe calibrated in 0.05 ml divisions, with an integral needle (e.g. Monoject— see Equipment Index) may also be found to be useful in some circumstances. Volumes of 1–50 μl can be delivered most accurately using a microlitre syringe (e.g. Hamilton or Terumo, see Equipment Index), but use can also be made of a 1 ml tuberculin syringe fitted in a micrometer screw gauge instrument. For this the syringe must be calibrated in the micrometer. This is done by replicate weighings of the amount of water or mercury which is delivered when a single turn of the micrometer screw has depressed the plunger a certain amount. The volume delivered, and consequently the number of turns required to deliver a particular volume, can then be calculated (e.g. 1 ml water weighs 1 g; 1 ml mercury weighs 13.6 g). Microprocessor-controlled mechanical syringe drivers are available that can simplify this process, but some models are rather cumbersome (see Equipment Index).

Hypodermic needles are manufactured either entirely in steel, or have a hub made of plastic. They can be obtained either in the opaque (most usual) or transparent form. Plastic hub needles are usually supplied sterile and individually packaged. Needles are made in different sizes and lengths and are usually described by a gauge number (e.g. 21G). The outside diameter of the needle tubing associated with a particular gauge number varies in different countries. The British Standard diameters for a selected range of gauge numbers used in this book are given in Table 2.

The point of a needle is bevelled, and this is specially shortened in the case of a Shick needle which is used for intradermal injection where the needle must travel in the skin for only a few millimetres. Hypodermic needles can also be obtained for specialised use, such as for i.v. infusion where the needle is attached to plastic tubing, and a plastic wing is provided to allow the needle to be anchored to the skin. These

Table 2. Selection of needle sizes of British Standard wire gauge.

Gauge No.	Diameter of wire (mm)
10	3.43
11	3.06
12	2.68
13	2.32
14	2.03
15	1.83
16	1.59
17	1.37
18	1.21
19	1.04
20	0.88
21	0.81
22	0.73
23	0.65
24	0.58
25	0.51
26	0.46
27	0.42
28	0.38
29	0.34
30	0.31

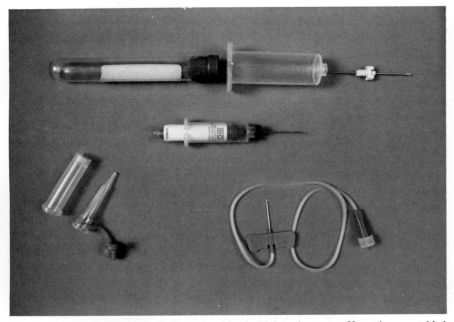

Fig. 1. Top: component parts of a Vacutainer blood withdrawal system; a Vacutainer assembled for insertion into the rat. Left: microvette blood collection tube and accompanying centrifuge tube. Right: butterfly or scalp vein infusion needle and catheter.

"Butterfly" infusion needle sets (Fig. 1) can be useful for intravenous injections, since the flexible catheter allows for some movement between the syringe and the needle without the needle becoming dislodged from the vein. If repeated injections of material are to be made over a short period of time, then it is preferable to use an indwelling catheter. These are available as either "over-the-needle" designs which have a needle placed inside the flexible catheter or as "through-the-needle" designs in which the catheter runs through the needle (Fig. 2) (see Equipment Index). The latter type enables a long catheter to be threaded through the introducer needle, but the diameter of catheter that can be placed in a vessel is smaller than with an over-the-needle system. The reader should consult the manufacturers of hospital and laboratory products if further information is required (see Equipment Index).

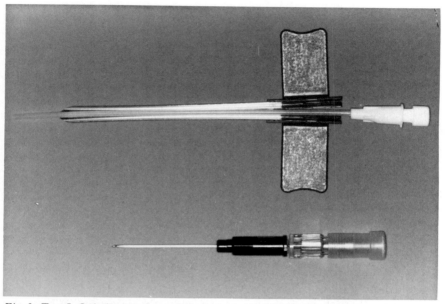

Fig. 2. Top: L-Cath through-the-needle catheter showing the needle split for removal. Bottom: An over-the-needle catheter.

2.1.2. Cannulae and Catheters

A cannula is a rigid tube made of glass or metal used to effect communication between a body cavity and the exterior. It is often used synonymously with the term catheter, but by convention the latter term is better used for a tube made of non-rigid material, e.g. plastic or rubber tubing. The tubing for cannulae and catheters is available with various internal diameters (i.d.) and outside diameters (o.d.), and may also be described using a French gauge number (see Table 3). Tubing for a variety of specialised, mainly medicinal uses can be obtained commercially (see Equipment Index).

2.1.3. How to Hold a Syringe and to Inject

The correct way to hold a syringe for injection is shown in Fig. 3. The syringe is held between the opposing sides of the first and second fingers, whose palmar surfaces lie against the flange of the syringe. The

Table 3. French gauge No. and equivalent external diameter.

French gauge No.	External diameter (mm)
2	0.63
3	1.02
4	1.34
5	1.65
6	2.10
7	2.30
8	2.60
9	3.00
10	3.30

thumb is initially placed on the top of the flange to give support while the needle enters the injection site. It is then transferred to the top of the plunger which it depresses to complete the injection. Where the injection is being made into a restricted space, such as a small vein, it is most important that an attempt should be made to steady the needle, for instance by pressing the needle against the thumb of the opposite hand while it is entering the vein, or resting the arm on a support, or steadying the hand by placing the side of the little finger on the table while making the injection.

The syringe can also be held, as in Fig. 4, for entry into any injection site; this allows greater stability of the syringe. However the hand has to be repositioned as in Fig. 3 before the plunger can be depressed, and this may entail unacceptable movement of the syringe, specially if the needle lies in a delicate vein. During the actual injection care must be taken to keep the needle in the same position at all times and not allow it to travel forwards or backwards while the plunger is being depressed. This pitfall, which is a consequence of the investigator having to shift his gaze from the needle to the fluid in the syringe, is often overlooked by the novice and requires a little practice to overcome.

Practice is also required to perfect the technique for entering blood vessels successfully, since the inexperienced investigator can completely pierce these in his initial attempts. Entry into any blood vessel is made at a shallow angle, with the bevel of the needle upwards, so that the needle point can "bite" into the vessel wall. Once it has penetrated, the

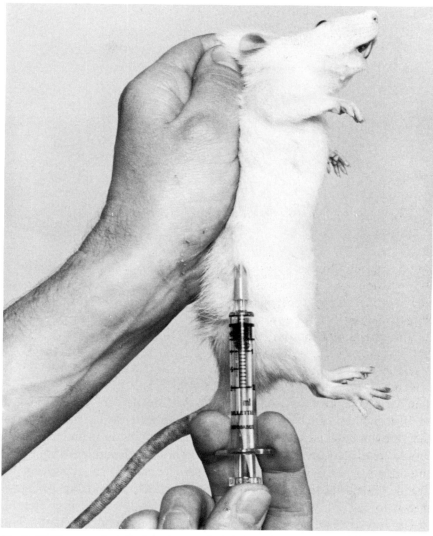

Fig. 3. Holding a syringe for injection, method 1; here the syringe is being used for a s.c. injection.

needle should be brought progressively more horizontal as the point is pushed slowly into the lumen. The needle can often be sighted while doing this in the larger blood vessels. These individual procedural steps will become incorporated into one fluid movement once experience has been gained.

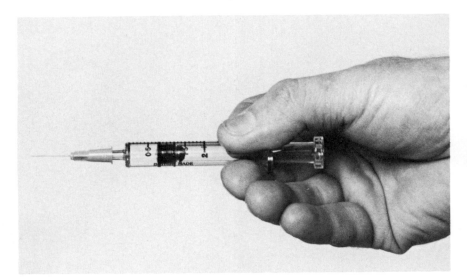

Fig. 4. Holding a syringe for injection, method 2.

2.2. Methods of Restraint

2.2.1. *Handling of Rats*

The large majority of laboratory rats are docile and accustomed to handling. They can be picked up and restrained easily, provided that they are approached and handled in a firm but gentle manner. The rat should be picked up by placing one hand around the animal's shoulders and lifting it clear of its cage. The handler should then position his thumb under the rat's mandible to prevent the animal from biting, as shown in Fig. 5. When handling larger rats (>200 g), it is preferable to support the animal's hindquarters with the operator's hand. This additional support also enables easy restraint to allow an assistant to carry out an intraperitoneal or intramuscular injection (Figs 12 and 13). If it is necessary to hold large animals singlehanded then the hindquarters can be immobilised if the tail is tucked between the fourth and fifth fingers (Fig. 6). It is important that only sufficient force is used to prevent the rat from escaping. Inexperienced research workers frequently tighten their grip on the animal if it struggles, and this can result in excessive pressure being applied to the trachea and thorax. The rat will

Fig. 5. Restraint of the rat. The handler's thumb is positioned under the animal's mandible to prevent it from biting.

then struggle even more violently, become cyanosed, and may also bite when it is released. To avoid this problem, if the rat cannot be restrained easily it should be released back into its cage. It can then be picked up again and handled gently for a short period before being fully restrained.

If the animal is agitated or appears aggressive, a safer approach is to remove the rat from its cage by grasping the base of its tail. It can then be transferred on to the top of the cage or on a surface on which its feet can get a grip. The natural instinct of the rat will be to pull away from the investigator, who must therefore hold it firmly. The rat will not pull only in one direction but will be constantly on the move. The investigator places the palm of his free (usually left) hand onto the rat's back. The thumb and forefinger are passed around the neck, with the thumb going in front of, or more usually behind one forelimb, to end up well under the lower jaw and the rat is then lifted clear of the cage-top and restrained as described above (Fig. 5). This method of restraint is probably the most comfortable for the investigator as well as the rat, and enables i.p. and i.m. injections to be carried out.

Fig. 6. Restraint of the rat. If it is necessary to hold large animals single-handed then the hindquarters can be immobilised if the tail is tucked between the fourth and fifth fingers.

Subcutaneous injections are less easy to administer when using this method of restraint, and administration of compounds by gavage in these circumstances is both difficult and hazardous to the animal. An alternative technique of restraint is to pick up the rat by the tail, as described above, and after placing the palm of the hand on the rat's back, the fingers gather up tightly the loose skin of the back and neck. In particular the thumb and forefinger must gather up the skin from around the ears. The head should be immobilised almost completely if the procedure is executed correctly (Fig. 7). This method may be strongly resented by some animals until they become accustomed to the

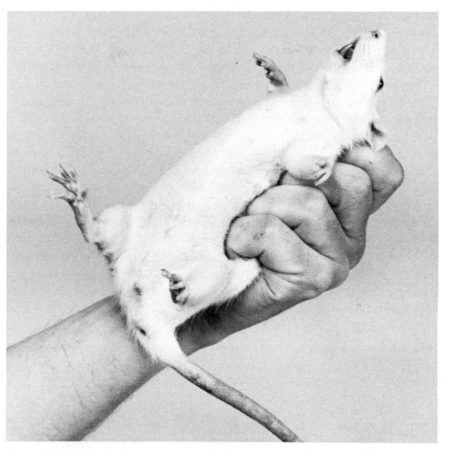

Fig. 7. Restraint of the rat by grasping the loose skin of the back and neck.

procedure. In addition the technique results in hand fatigue if many animals are to be held or if large rats are involved. The method is particularly well suited to administration by gavage of material because the head is immobilised firmly.

2.2.2. *Restraint Devices*

If the animal proves difficult to catch and to handle, then it may be preferable to use a restraining device rather than subject it to unnecessary stress caused by prolonged handling. Use of a restraining device may also be necessary if an assistant is unavailable. The simplest restraint device is to roll up the animal in a small towel which is secured with safety pins or artery forceps. That part of the body required for an injection, e.g. the tail or foot, is left sticking out. A second, simple technique is to use a polyethylene restrainer of the type shown in Fig. 8 (see Equipment Index). This simple device was originally designed to provide secure restraint prior to euthanasia by decapitation using a guillotine, but it can be used to provide restraint for a wide variety of purposes. Both subcutaneous and intraperitoneal injections can be carried out through the polyethylene, and it is especially useful if these procedures need to be carried out without the help of an assistant.

An alternative technique is to prepare a tube made of plastic (e.g. domestic drain piping) or of metal sheeting with an internal diameter of 4–5.5 cm depending on the size of the rat. The length of the tube should be such that the rat just fits inside with its tail outside. If the tube is provided with a perforated metal plate fitting into slots to adjust the length, rats of various sizes can be accommodated. One end of the tube should be capped with perforated metal or wire gauze to allow free

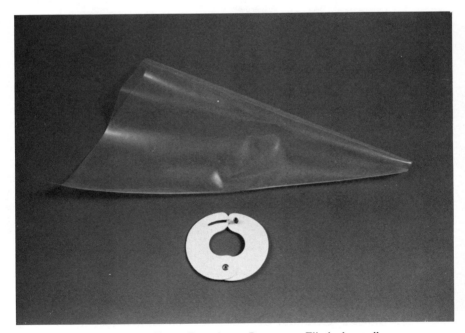

Fig. 8. Restraint apparatus. Top: a Decapicone, Bottom: an Elizabethan collar.

entry of air (if the slot arrangement is not employed), while the other end is stoppered with a rubber bung with a V cut out of it. This allows the tail of the rat to be held with restricted movement for injections into the appropriate site or vein (Fig. 9). For use the tube can be held in a retort stand and clamp, and placed at any angle.

Many other types of restraining apparatus have been described which are used during injection procedures, and some commercially available ones are shown in Fig. 10a, b and c (see also Abildgaard, 1964, and Equipment Index).

2.3. Subcutaneous Injection

The injection is usually made under the skin of the back and sides. The needle should be passed through the skin in an anterior direction and at a shallow angle to the skin surface (see Fig. 3). An alternative approach that can be adopted with a docile rat is to restrain the animal as shown in Fig. 11 and inject into the skin overlying the neck. When in position, the tip of the needle should be moved up and down to reveal its whereabouts and also to ascertain that the needle is truly s.c. If the tip cannot be discerned then the needle could be i.p. or i.m. and must be withdrawn slightly to lie subcutaneously. A successful s.c. injection made on the back or sides results in the formation of a bleb

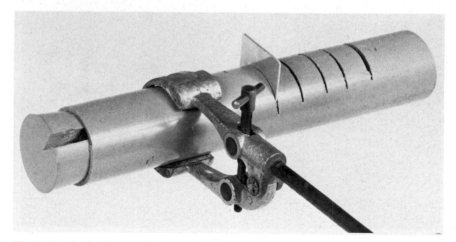

Fig. 9. Plastic pipe for restraining a rat; the slots allow different sized rats to be accommodated.

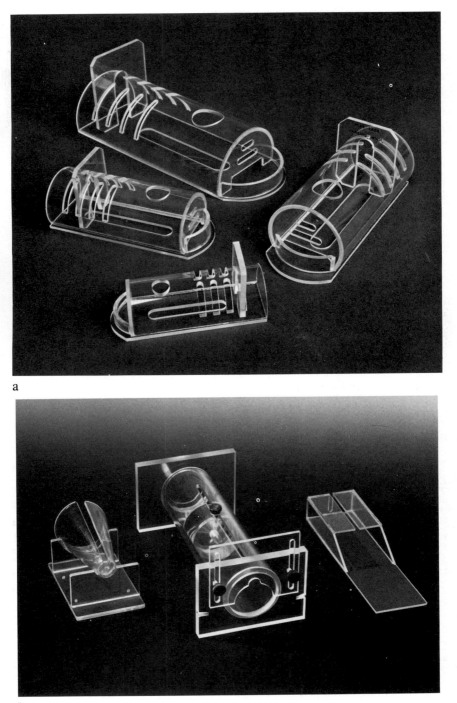

a

b

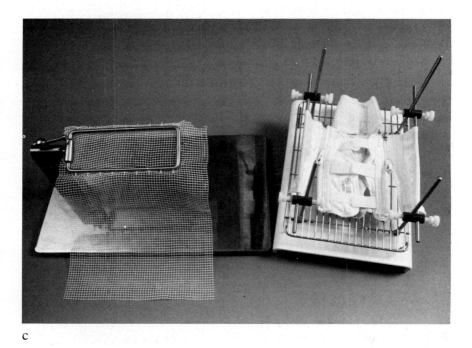

c

Fig. 10. Commercially available rat restrainers.

during discharge of about 0.5 ml or more of solution from the syringe. Ideally, the whole of the needle shaft should be s.c. as this ensures the minimum leakback of solution. When injecting a large volume subcutaneously (e.g. >2 ml), leakback and hence loss of fluid can be minimised further by changing the needle path after the needle has been pushed in half way. Irrespective of the size of the animal, the smallest size of hypodermic needle should be used compatible with the type of material being injected. A 25G needle for example is suitable for non-viscous aqueous solutions whereas a 23G or 21G needle would have to be used for viscous solutions or for suspensions.

2.4. Intraperitoneal Injection

The injection is made into the lower left quadrant of the abdomen. In this area there are no vital organs except for small intestine. In contrast,

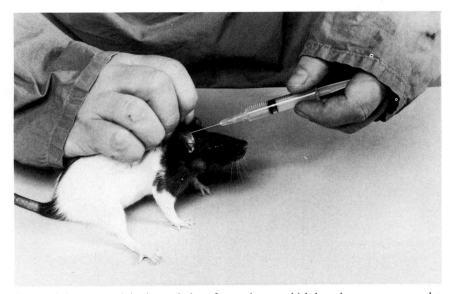

Fig. 11. Subcutaneous injection technique for use in rats which have become accustomed to handling.

the lower right quadrant contains much of the large caecum, and the upper abdomen is a hazardous area to inject into because the liver, stomach and spleen are situated there. When holding a small (<200 g) rat, the animal can be adequately restrained to allow the injection to be made by the operator. When administering compounds to larger animals, it is advisable to have an assistant. The hindquarters and tail are restrained by the assistant, and the operator extends one of the animal's hind legs and carries out the injection (Fig. 12). For the injection, the tip of the needle should first be inserted subcutaneously, though this step is unnecessary once the technique has been mastered, and a final short thrust is then made through the abdominal muscles, holding the syringe nearly vertical. It is necessary to insert only the tip of the needle into the peritoneal cavity otherwise the intestine can be punctured. If the injection is made into the intestine in error, fluid will often be seen issuing from the rectum immediately after the injection. However, this must not be confused with the animal urinating.

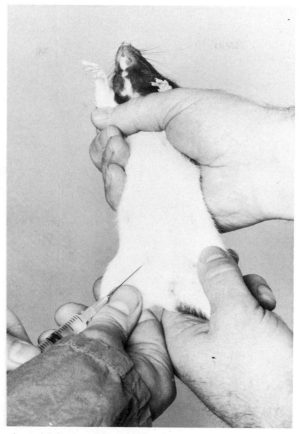

Fig. 12. Intraperitoneal injection. The injection is made into the lower left quadrant of the abdomen.

2.5. Intramuscular Injection

The usual site for this route is into the muscles of the hind limb. Either the biceps femoris, the semitendinosus and gluteus maximus muscles which make up the posterior aspect of the thigh and rump, or the quadriceps muscle group on the anterior thigh are used (Figs 13 and 14). The needle should not be inserted too deeply otherwise bone may be encountered. It is not easy to be sure that the needle is truly i.m., however it should not be possible to feel the point of the needle through the skin if it is indeed in the muscle and not s.c. Sometimes an i.m.

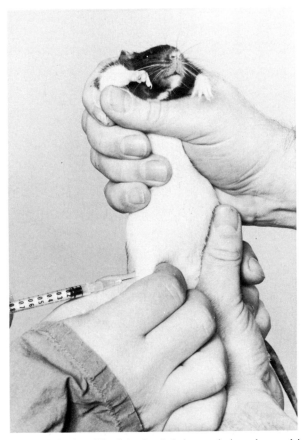

Fig. 13. Intramuscular injection. The injection is being made into the quadriceps muscle on the anterior aspect of the thigh.

injection will fail, even though the needle is felt to be in the muscle mass, because it actually lies in one of the fascial planes.

2.6. Intradermal Injection

The most usual sites are the skin over the back and abdomen, or over the ventral surface of the hind feet. The hair should first be removed with clippers and a depilatory can be used if required. The investigator should note that depilatories are chemical substances which could

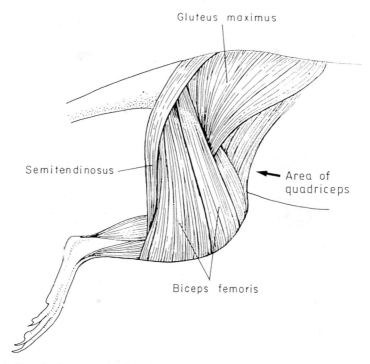

Gluteus maximus

Semitendinosus

← Area of
quadriceps

Biceps femoris

Fig. 14. Sites for intramuscular injection.

interfere in the experimental study. Their use therefore should be considered carefully, and the animal prepared the day before the experiment if possible. A special short-bevelled hypodermic needle (Shick needle) is available for intradermal injections, however, an ordinary short-length 25G or 26G needle will often suffice.

For the injection, the needle is held bevel down almost flush with the surface, and is pushed into the skin for 2–3 mm before the injection is made. Since it is easy for the needle to travel subcutaneously rather than intradermally, it should be noted that in contrast to a s.c. injection, there is considerable resistance to the passage of the needle when it is being inserted into the dermis. Since only a very small quantity of material can be deposited intradermally (50–100 μl) a successful i.d. injection raises a bleb, whereas a s.c. injection of such a small volume would not do so. If a coloured substance or emulsion is injected this will be visible in the skin (Fig. 15), whereas if the injection is made

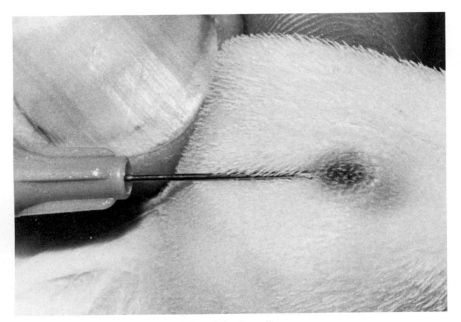

Fig. 15. Intradermal injection; the injected dye is easily seen.

s.c. in error the material would probably not be seen. These differences should be looked for when carrying out an intradermal injection.

2.7. Topical Application

The usual site for topical administration is the skin of the back. The hair must be removed with clippers or a depilatory (but see 2.6). The substance to be administered is then applied in a volatile solvent or a cream to the denuded skin, with a dropper, a cottonwool applicator, a brush, or it may be smeared on. For administration to a large number of animals, a syringe fitted with a blunted needle and connected to a variable metering device (e.g. the Repette—see Equipment Index) will prove very useful for soluble materials or suspensions.

Some thought should be given to the state of the hair cycle when administering substances topically. It is usual to use animals not showing active hair growth at the site of administration since in these there is quicker and better absorption of the materials. Subjectively, hair growth is usually evident if a short stubble or fine "down" is present which cannot be clipped away. When clipping hair for topical administration, the clipper blades should be cleaned of any oil or grease since, being extraneous chemical substances, they may affect the response to the chemical material under investigation.

2.8. Intragastric Administration by Gavage

2.8.1. Administration in the Adult Rat

The administration is carried out using a 15G or 16G curved or straight blunt-ended needle cannula about 11 cm long (Fig. 16) (see Equipment Index). Administration is safer and more sure if the tip of the needle is "bulbed". The diameter of the bulb will depend on the size of the rat to be gavaged. Leslie and Conybeare (1988) recommend a range of sizes (Table 4) which are calculated to prevent the regurgitation of fluid up past the bulb which, if it occurred, might enter the trachea and prove fatal. If desired the bulb can be produced by winding around the needle tip some silver solder. Alternatively an acorn-shaped tip machined

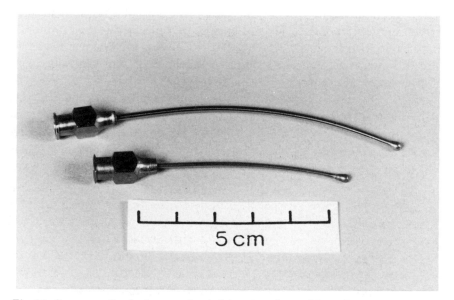

Fig. 16. Gavage needles for intragastric administration of materials to the rat.

Table 4. Size of bulb for gavage needles for rats of different body weights.

Rat weight (g)	Diameter of bulb (mm)
Up to 75	2.0
75–90	2.5
90–140	3.0
140–180	3.5
> 180	4.0

from brass rod, or a steel ball bearing, both with a suitable central hole drilled in them through which the tip of the needle can just pass, may be stuck or soldered onto the needle tip (Fig. 17). The bulb helps to prevent the needle from damaging the oesophagus and also from passing through the glottal opening into the trachea.

Before the administration, a note should be made of approximately how much of the needle would protrude from the mouth after passing into the stomach (measure this on the outside of the rat). If much more than this amount protrudes during the administration, then intubation of the trachea in error would be indicated and the needle must be removed for a fresh try. For the administration, the rat is held very firmly by the skin of the neck and back so that the head is kept immobile and in line with the back. The needle, attached to the syringe, is passed into the mouth as far to one side as possible (N.B. not centrally), and after locating the entry to the oesophagus is pushed gently into the stomach (Fig. 18). The passage of the needle may be obstructed in two places, viz. at the back of the mouth and at the sphincter to the entrance of the stomach. Manipulation of the syringe to produce a gentle thrusting movement combined with a gentle backward and forward movement will often overcome any difficulties. Occasionally the needle may have

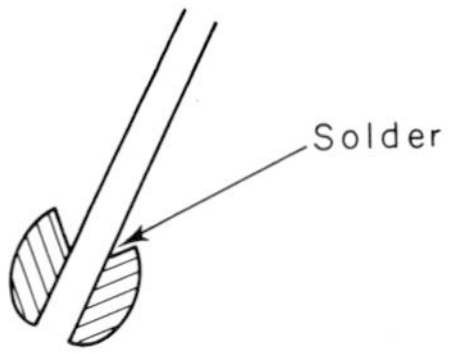

Fig. 17. Preparing a bulbed needle for safe intragastric administration.

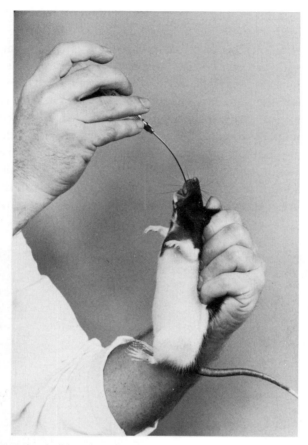

Fig. 18. Intragastric administration using a rigid dosing needle.

to be removed and inserted afresh. Discharge from the needle can be carried out fairly rapidly after a slow initial discharge to ascertain that the needle is truly intragastric. Some pointers to the fact that the needle is intragastric and not tracheal are: the rat may continue to "squeak" while the injection is being made, this would not happen if fluid was getting into the lungs; the rat may show swallowing movements and should not struggle unduly during the administration.

If struggling is severe or if the injection solution appears in the mouth or at the nose, then there is a good chance that the injection is being made in error into the lungs. The administration must be stopped immediately. If lung dosing is suspected, then the animal should be humanely killed, since severe pneumonia is likely to occur.

An alternative method of oral administration is by intubation with a rubber or plastic catheter (French gauge 8). To carry out the procedure in the conscious rat a gag has usually to be employed to prevent the rat biting through the catheter, though with docile rats the gag can be omitted. The oval-shaped gag is made from wood or plastic with a central diameter of about 2 cm and a central hole of 5 mm diameter (see Fig. 19 and Equipment Index). The catheter should have an external diameter of about 2.6 mm, and is fitted onto a syringe containing the liquid to be administered. The catheter should be filled with the injection material before insertion so that accurate dosage can be achieved.

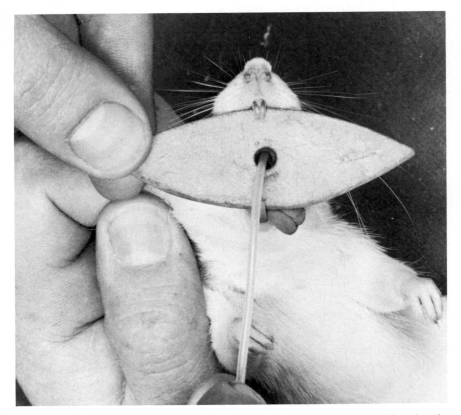

Fig. 19. Intragastric administration using a gag and a flexible dosing catheter. Note that the tongue is brought forward and held out by the pressure of the gag.

The rat is held by the skin of the back and neck, preferably by an assistant who also carries out the gagging procedure. With practice, however, the method can be carried out in its entirety by one person. One edge of the gag is then placed in the mouth as far back into the angle as possible, the flat part of the gag resting on the tongue. The gag is now turned in a scooping motion so that it lies vertically behind the incisor teeth. The scooping motion forces the tongue forward out of the mouth and therefore out of the way, and it is held like this during the administration. The catheter is now passed through the hole in the gag and into the stomach. The considerations given in the first method above to ensuring that the catheter lies in the stomach and not in the lungs also apply here. It should be noted that the conscious rat makes swallowing motions which help to guide the oral catheter into the oesophagus and consequently gastric intubation of the anaesthetised rat may be more difficult by this method.

In the rat it is possible to administer gelatine capsules into the stomach to feed solid materials. To do this, a stainless steel cup (e.g. 4 mm) with a hole in the bottom to accommodate the tip of a needle, is soldered onto a slightly bent 16G blunt-ended needle cannula (Fig. 20).

Alternative cup sizes can be prepared to fit different sized capsules. Capsule administering devices are also available commercially in some sizes. Also commercially available are capsules, some as small as 2.4 mm $\times$ 7.3 mm (see Equipment Index). The filled capsule is placed firmly into the capsule holder with the capsule cover facing the interior. The rat is held as for intragastric administration. The capsule, cup and cannula attached to a syringe with the plunger partly withdrawn, are inserted into the mouth, and the cannula is then pushed along the palate to the oropharynx. At a depth of about 2.5 cm the capsule is ejected by air which is forced through the cannula and the cup by depressing the syringe plunger, and the feeder is quickly pulled out. The rat's

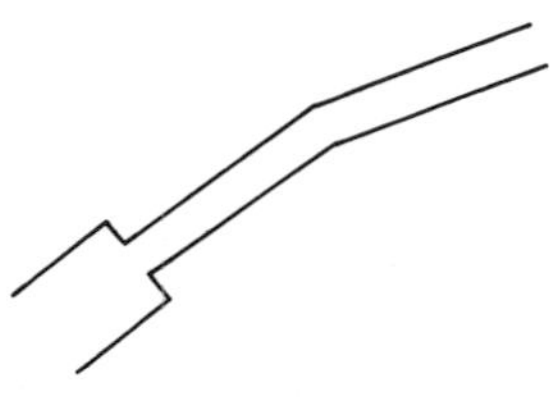

Fig. 20. A needle modified for holding gelatin capsules.

mouth may have to be held shut until it swallows. Up to 150 mg of powdered material can be fed by this means.

2.8.2. Administration in the Young Rat

Rats of weaning age (3 weeks) are dosed using a 19G or 21G needle 40–50 mm long. The needle should be blunted and bent by 20–30°, and the tip can be "bulbed" if required. The rest of the procedure is as for 2.8.1 (first method).

2.8.3. Administration in the Neonatal Rat

Rats aged 0–14 days are intubated using silicone rubber tubing, 0.33 mm i.d. and 0.64 mm o.d. Polyethylene tubing should not be used as it may perforate the oesophagus. The rat should be held gently either by the skin of the neck and back, or by sliding its hindquarters into a conical centrifuge tube. A drop of water is placed on its muzzle to induce swallowing and, using forceps to hold the tubing, the catheter filled with the material to be injected and connected to a syringe, is fed into the mouth and the animal is allowed to swallow a small piece at a time. The tube must not be forced in. Approximately 5 cm of tubing is needed to pass into the stomach, but a little more does not harm the animal. When injecting, the animal must be watched carefully to see if fluid is regurgitated or if there is any respiratory distress, both of which indicate tracheal intubation in error. About 0.1–0.2 ml can be administered comfortably.

2.9. Intra-articular Injection (into the Knee Joint)

The needle to carry out the injection is prepared as follows. A piece of polyvinylchloride (e.g. Tygon) or polyethylene tubing i.d. 0.025 mm, o.d. at least 0.076 mm, and 5 mm long is soaked for a few minutes in dichloromethane to soften and swell it. It is then slipped over a 26G $\times$ $\frac{3}{8}''$ needle and pushed up to the hub. The tubing contracts and attaches firmly to the needle. This acts as a stop to prevent the needle going too far into the knee joint. The needle is now sterilised either with disinfectant or by ethylene oxide (see 15.4.3). For the injection, the rat is anaesthetised and placed on a table on its back. The leg over the region of the knee is swabbed with antiseptic and then stretched out straight, providing gentle traction. The space between the knee joint is palpated with a finger wearing sterile gloves to ascertain its position. With the first finger of the hand which is holding the leg placed under the knee to support it, the needle, held vertically, is inserted through the knee tendon into the space between the knee joint (Fig. 21). Up to 15 μl of fluid can be injected comfortably.

2.10. Subplantar Injection

The rat is restrained either manually by an assistant or in a suitable restrainer. The sole (plantar aspect) of the back foot is used for the injection. The injection is made subcutaneously with the point of insertion of the needle being about half-way along the sole (Fig. 22). A 26G × $\frac{3}{8}''$ needle attached to a 1 ml syringe, or a 0.5 ml insulin syringe can be used effectively to inject up to 100 μl of fluid. Slight pressure is placed on the point of needle insertion for a few seconds after the needle is withdrawn.

Fig. 21. Intra-articular injection of material into the stifle (knee) joint.

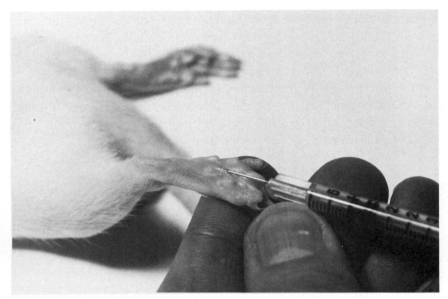

Fig. 22. Subplantar administration.

2.11. Intratracheal Administration

The apparatus required for the instillation of material into the lung of
an adult rat is shown in use in Fig. 23. The apparatus consists of a
12.5 × 5 cm piece of Perspex attached upright to a heavy wooden base
and having its upper half inclined by about 20° towards the investigator.
The total height of the apparatus is 30 cm. Rats treated with atropine
sulphate to prevent salivation (see Table 7) are anaesthetised with
methoxyflurane or halothane and suspended by their forelimbs with
loops of string. Suspension is rapidly effected by fastening the loops
round the limbs by a sliding piece of polythene tubing (see 15.2.1 and
Fig. 76), provided that the two free ends of the loops then pass through
a hole in the Perspex small enough to grip the double strand of string
without tying. The head of the rat is pressed against the Perspex by a
rubber band which passes round it and under the upper incisors. For
intubation the rat's tongue is pulled out and to one side by an assistant
using smooth blunt forceps, and the inside of the back of the mouth is
illuminated by the investigator with a laryngoscope, of the type used
for children, with the speculum removed. Because the area is well

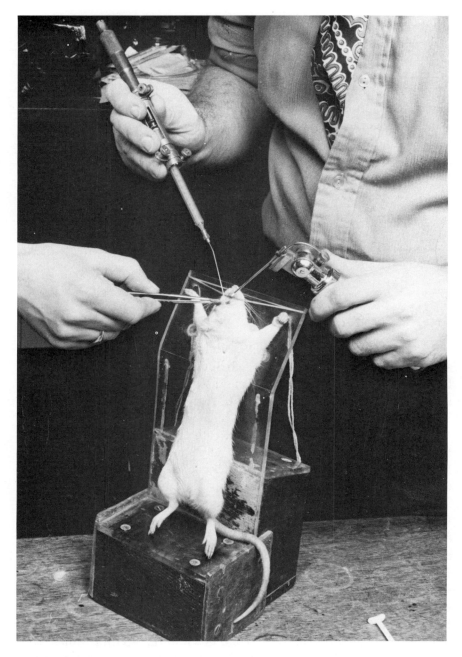

Fig. 23. Intratracheal administration; note the modified paediatric laryngoscope and the tuberculin syringe held in a micrometer screw gauge.

illuminated, it is a simple matter to pass a slightly curved, blunted hypodermic needle (22G, 50 cm long), attached to a tuberculin syringe held in a micrometer screw gauge, between the vocal chords and into the trachea for approximately 2.5 cm. Alternatively a piece of small bore polyethylene tubing attached to a needle can be used. Up to 40 μl of material can then be placed at the lower end of the trachea by the assistant manipulating the micrometer screw, the micrometer having been calibrated previously for use with a particular syringe (see 2.1.1). After depositing the material in the trachea, some of it will pass into the lungs. Since anaesthesia is not continued while the rat is on the holding apparatus, the entire procedure must be carried out rapidly before the animal starts to recover. In this way, it is possible to intubate rats at the rate of one a minute. Thet (1983) and Costa *et al.* (1986) have described various methods of intubation of the trachea using specially designed apparatus (see Fig. 70).

2.12. Intravesicular Administration (into the Bladder)

Catheterisation of the bladder is easiest to accomplish in the female, and very difficult in the male rat. The procedure will only be described for the female rat. Small 7.5 cm lengths of polyethylene tubing (either i.d. 0.28 mm, o.d. 0.61 mm, or i.d. 0.4 mm, o.d. 0.8 mm) are sterilised by autoclaving or by immersion in antiseptic followed by washing in sterile saline. Alternatively the flexible catheter from a suitable size over-the-needle i.v. catheter can be used. The length of the catheter should be such that only about 10 mm is sticking out after catheterisation, so that any dosage error due to the dead space in the catheter is minimised. The rat is anaesthetised and held in a supine position, with the head towards the investigator and the base of the tail between the dorsal surface of the first finger and the ventral surface of the middle finger of the left hand. The thumb should be placed on the abdomen about 1 cm anterior to the urethral opening and pulled backwards slightly to stretch the urethra (Fig. 24). The sterile catheter handled with sterile gloves is now inserted into the urethral opening and pushed along as far as the vaginal opening. The position of the catheter can be discerned quite easily if the tip is moved up and down in the urethra, which runs just under the surface of the skin. The free end of the catheter should now be pointing towards the investigator. The thumb of the left hand

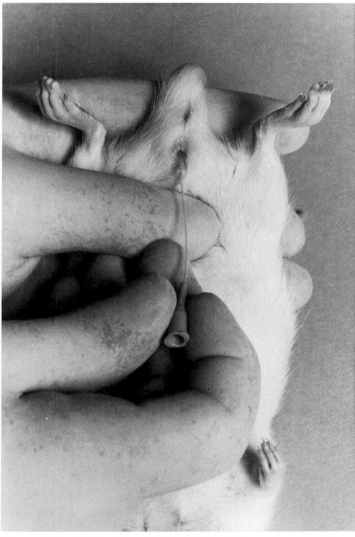

Fig. 24. Intravesicular administration; initial intubation of the urethra; note the way in which the rat's tail is held between the first and second fingers, the head hangs downwards.

which is moved up to the base of the urethra should now be pushed forwards so that the end of the catheter sticking out of the urethra is made to point away from the investigator (Fig. 25). This manoeuvres the catheter round the pubic symphysis after which it is pushed forward

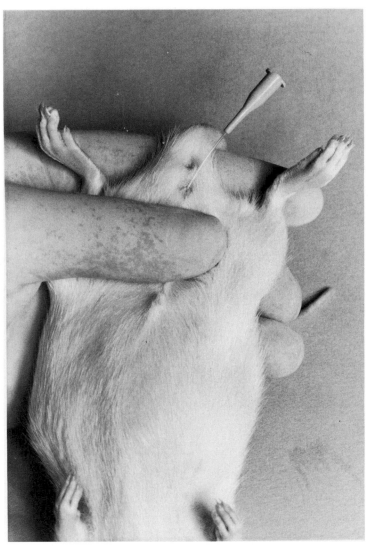

Fig. 25. Intravesicular administration; the position of the catheter is reversed in order to get round the pubic symphysis.

into the bladder. Some gentle manipulation and turning of the catheter may be necessary for this to be accomplished successfully. In some cases intubation may be unsuccessful, the catheter should therefore be removed and fresh attempts made. Any infection in the urethra will make intubation difficult. The syringe containing the solution to be administered is inserted into the catheter via a 30G or 25G needle depending on the size of catheter. The administration is made slowly taking care that no solution comes out of the urethra. A volume of about 0.15 ml is held easily in the bladder.

Since administration into a full bladder will result in considerable dilution of the administered material, or more likely in the solution being voided and therefore lost along with the urine immediately after the catheter is removed, intubation should be carried out in animals with empty bladders. Mere handling of the animals when inducing anaesthesia initially often accomplishes this, but the abdomen over the area of the bladder should be compressed several times to make sure.

Repeated intubations can be made by this route, but infection is a problem, and severe infection is often seen as a white encrustation around the urethral opening. Also bladder stones (calculi) may form some months after one or more intravesicular injections.

The use of catheters made from polyethylene has been stated by some investigators to be contraindicated because of their tendency to inflict mucosal damage. The use of special polyurethane ureteral catheters with or without coudé (curved) tips has been advocated. The coudé catheters in particular should facilitate intubation where getting round the pubic symphysis proves to be a problem. These catheters can be obtained with outer diameters as low as 1 mm (size 3 French, see Equipment Index).

2.13. Intracerebral Injection

Injection of materials into precise parts of the brain can be carried out using a small animal stereotaxic apparatus (Fig. 26), in conjunction with coordinates shown in a stereotaxic atlas of the rat brain (see Pellegrino and Cushmann, 1967; Caulfield et al., 1983). However, a simple method which allows injection into grossly defined areas of the brain is as follows. Adult rats are anaesthetised and placed on their ventral surface. A midline incision into the skin is made with a scalpel using aseptic technique (see Section 15.5.1) from between the eyes to the level of the ears, and the junction of the coronal, sagittal and transverse sutures exposed (see Fig. 27). Entry through the skull is effected by using a dental burr of about 1 mm diameter held in a dental drill cap (see

Fig. 26. Stereotaxic instrument for small animal. (a) Ear bars.

Equipment Index). The actual position of the drill hole will be dependent on the part of the brain to be injected. For an injection into the substance of the frontal lobe the hole must be situated about 1 mm in front of the transverse blood sinus, easily seen through the skull, and about 1–2 mm lateral to the sagittal sinus (Fig. 27). For injection into the lateral ventricle the hole is placed 1.5–2 mm lateral and posterior to the junction of the sagittal and coronal sutures (Fig. 27).

The injection is made using a hypodermic needle of size between 21G and 27G, attached either to a microlitre syringe or to polyethylene tubing which is attached to a tuberculin syringe held in a micrometer screw gauge, calibrated for that particular syringe (see 2.1.1). For

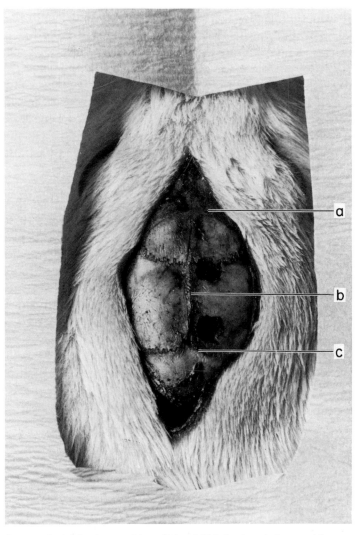

Fig. 27. Intracerebral injection; position of the drill holes in relation to: (a) coronal suture and underlying sinus, (b) sagittal suture and underlying sinus, (c) transverse suture over transverse sinus.

injection into the frontal lobe about 3.7 mm of the needle tip needs to be pushed into the brain. For injection into the lateral ventricle, with the needle held vertically, 3.5–4 mm of the tip needs to be introduced into the brain. The procedures are simplified if a piece of fitted tubing is slipped over the needle such that only the required length of needle is exposed. The tubing then acts as a stop against the skull. Up to 30 μl of fluid can be injected into the ventricle without leak-back, provided the needle is left in place for about 5 seconds after the injection and that light pressure is placed on the needle site for a few seconds after withdrawal of the needle. Although 30 μl can also be expelled into the frontal lobe or any other part of the brain substance, some leak-back, together with escape of a little cerebrospinal fluid, will be encountered. It is advisable, therefore, to inject volumes smaller than 30 μl into these sites.

Although figures have been given for the lengths of the needles necessary for the injection into two brain sites, in practice the actual lengths will vary with age and body weight (Brakkee et al., 1979). It is imperative, therefore, to check that the solution is being delivered to the desired site. This can be done by injecting ink or a histological stain, and then carefully dissecting the brain to determine the position of the dye.

After an intracerebral injection the skin incision is closed with sutures or skin clips. In most cases rats survive the injection very well without overt signs of neurological dysfunction.

In young rats of weaning age or younger the skull is very thin, and entry can easily be effected without surgery. After anaesthetising the rat, a hypodermic needle can be forced directly through the skin and bone and into the brain with a twisting movement.

2.14. Intraorgan Injection

The injection of materials into an organ presents no unusual difficulties. The smallest needle available should be used, and the injection carried out very slowly to minimise leak-back. The amount that can be injected is very small and in the region of 10–50 μl depending on the size of the organ.

2.15. Intravenous Injection

A number of different veins are available for i.v. injection in the rat. Which vein to use will depend on the preference of the investigator and whether that vein is available.

Some general points on i.v. injection should be noted. Where injection is into an exposed vein the needle can be judged to be in the vein when it slides freely backwards and forwards, and when the tip of the needle can be seen when raised and cannot be seen when lowered, i.e. reflecting blood passing under and over the needle tip respectively. In large veins, whether exposed or not, it should be possible to withdraw a little blood into the syringe if the needle is correctly in the vein. If the position of the needle is in doubt then injection of a very small amount of the solution will raise a bleb if the needle is not in the vein but in the tissue surrounding it. In this case a fresh attempt must be made, or the needle moved in the surrounding tissue in such a way that it then enters the vein (or re-enters the vein if it has completely traversed it). With some exposed veins an attempt to enter the vein causes the vein to move forwards and offer a reduced resistance to the needle. This is because the vein has little in the way of connective tissue holding it in place. One consequence of this forward movement is that the investigator tries to enter the vein more forcibly, and upon entry the vein springs backwards, and together with the extra force used, this results in the vein being completely transfixed. To overcome this the vein must be prevented from moving forwards. This is accomplished by straddling the vein with two fingers a little way behind the intended site of entry into the vein. In this way the vein is "held down" without obstructing the blood flow. The needle can then enter the vein after first passing through the gap between the two fingers, or after passing close to one side (see Fig. 50).

The investigator should note that if a vein is to be used for serial injections at short time intervals, the first injection should be made in the vein at a distal site in relation to the heart, subsequent injections being placed progressively more proximal. This is because the blood is flowing towards the heart, and if for a particular vein the most proximal site to the heart were to be used, then if the injection damages and blocks the vein at this point, the distal portion of the vein would not be available for subsequent injections.

A final important consideration in selecting a site for venepuncture is the consequence of inadvertent extravascular administration of material, or the production of intravascular thrombosis or thrombophlebitis. Although such complications must always be considered undesirable, the clinical problems and the distress caused to the rat may vary considerably depending upon the vein involved. In general, bruising and even occlusion of one tail vein is well tolerated, and the extensive collateral circulation that is present prevents the development of oedema. Occlusion or damage to the sublingual veins and penile vein is much less well tolerated, and it is recommended that these sites are used only under exceptional circumstances.

The lateral tail vein is probably the easiest method of intravenous injection in which to become proficient. It is also well tolerated by the rat, and the consequences of technical failures are less serious than when injection is made into other veins. The dorsal metatarsal vein is also relatively straightforward to use, although the technique requires the use of an assistant.

2.15.1. Injection into the Tail Vein

2.15.1.1. By needle. There are four equatorially placed blood vessels in the rat tail, the two lateral ones being the veins. The rat is first suitably restrained (see 2.2.2). Tail vein injection is often considered a difficult technique, particularly in dark skinned rats and also because the skin is tough and hinders penetration. However, once the technique is learned, its simplicity will be appreciated. The secret of successful injection of the tail vein is to dilate the veins by one of the following procedures. (i) Place a tourniquet around the base of the tail. This can be done simply by placing a piece of string round the tail and twisting it tight. A rather better way is to pass a piece of thread through the rubber end of the plunger of a 2–10 ml syringe and out through the syringe point and tie to form a loop. The tail is placed through the loop and the plunger withdrawn forming a tourniquet round the tail (Fig. 28) (Minasian, 1980). The tourniquet should be released just before the injection is made. (ii) Immerse the tail in hot water just bearable to the hand (about 40°C) for 1–2 minutes. (iii) Place the rat in an incubator, oven or "hot box" maintained at 37°C, for 10–15 minutes. These devices are available commercially (see Equipment Index) or one can be made from a cardboard or wooden box containing a thermometer, over which is placed a 100 W light bulb or an infrared heat lamp (see Equipment Index). If an infrared lamp is used rats should be exposed to it for only 2–3 minutes to prevent skin burn. Providing that the box is prewarmed before putting the animals in, this time period is sufficient to cause adequate vasodilation of the tail veins.

When the rat is removed from the "hot box", the tail should be plunged immediately into water maintained at 40°C for about 30 seconds.

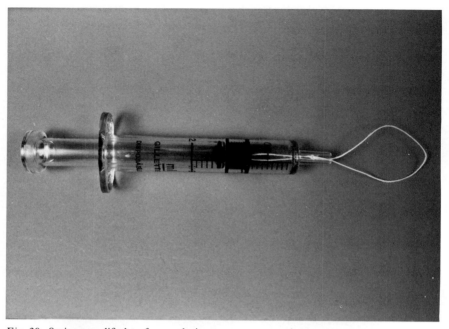

Fig. 28. Syringe modified to form a device to use as a tourniquet.

The tail is removed and dried thoroughly with a paper towel. This both further dilates the veins and more importantly, cleans and softens the skin which facilitates injection.

The injection is carried out using a 23–25G needle. The lateral veins are immediately beneath the skin and must be entered at a very shallow angle, almost parallel to the vein. The tail should be bent down with the left hand while the vein is being entered at the point of the bend with the needle and syringe held in the right hand (Fig. 29).

It is worthwhile noting that using needles with a translucent hub makes it easier to verify that the vessel has been entered as blood can be easily seen to be flowing back. Occasionally the plunger may need to be drawn back. If it cannot be verified by backflow of blood that the vein has been entered, a small amount of fluid should be injected. If a bleb appears then entry has not been successful and the investigator will have to try again using a more anterior section of the tail.

2.15.1.2. By catheter. In some circumstances it may be necessary to infuse material over a short period, or carry out a series of injections at short intervals. A rapid and simple method for maintaining vascular

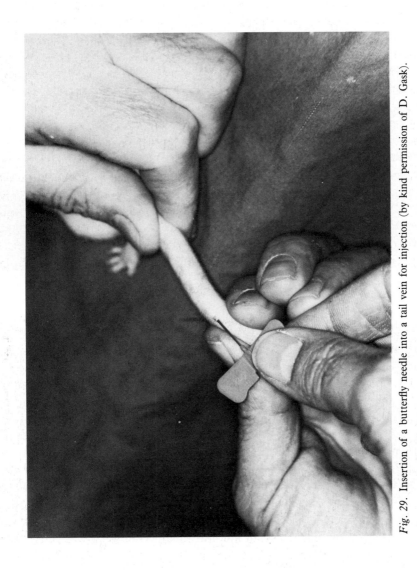

Fig. 29. Insertion of a butterfly needle into a tail vein for injection (by kind permission of D. Gask).

access is to use a 23 or 24G over-the-needle catheter (Figs 2 and 30, see Equipment Index). The catheter and its needle introducer are placed in the vein in the same manner as for intravenous injection. After positioning the tip of the needle and catheter in the vein, the needle is partially withdrawn whilst holding the catheter firmly in place. The catheter should fill with blood as the introducer is withdrawn. If this does not occur, carefully slide the introducer back into position and reposition the introducer needle tip and catheter. Once it has been successfully positioned the catheter should be advanced into the vein, still leaving the introducer occupying a little over half the catheter. This prevents the catheter flexing and kinking as it is advanced. The needle can then be withdrawn completely. It is then often useful to attach a transparent silicone rubber injection port (see Equipment Index) to the catheter hub to prevent further flow of blood. The catheter is taped to the tail and the injection is made through the port taking into account the "dead volume" of the catheter and port. A "heparin lock" will have to be introduced into the catheter if it is to be used for serial injections to prevent clotting of the blood at the catheter tip (see 22.2). Animals

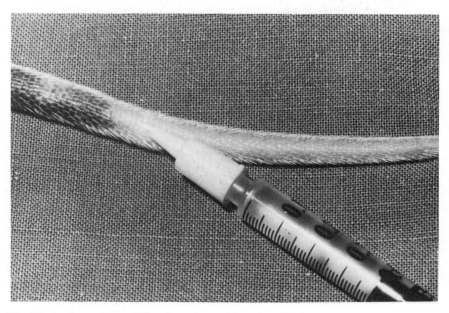

Fig. 30. An "over-the-needle" catheter inserted into the tail vein.

used for serial injections will either have to be restrained (see 2.2 and 17) or the catheter protected by tape or other means.

When it is required to administer substances via the tail vein either intermittently or continuously by infusion over a longer period spanning say, several hours, this can also be done by inserting a cannula in the tail vein of a conscious animal. When maintaining a cannula for this longer period, it may be preferable to insert a longer catheter into the vein. For the technique, the animal is warmed in a hot box (see 2.15.1.1) and then placed in a restrainer or held by an assistant. The infusion apparatus consists of a sterile disposable L-cath peel-away cannula 20G × 19 mm (Fig. 2, see Equipment Index) with its "through-the-needle" polyurethane catheter removed. This is replaced with a sterile single use intravenous catheter o.d. 0.63 mm and length 30 cm with an integral Luer hub fitting (Portex Ltd, cat. no. 200/300/010). The catheter is filled with saline and marked with an indelible pen to indicate the length to be inserted into the vein. If it is necessary, the veins can be further dilated by placing the tail in hand-hot (40°C) water for 2 minutes. The needle is inserted at a shallow angle into the vein and a small quantity of blood withdrawn or there is a show of blood to verify correct placement, before the cannula is pushed in the required length. The needle is withdrawn, split and removed leaving the cannula in place. Because of the special characteristics of the needle which allow it to be split in half, it can be removed from the catheter in spite of the other end being attached to a Luer fitting. The Luer fitting is then filled with a little saline to remove air and attached to the infusion line and syringe of the infusion pump. The catheter is secured by taping it to the tail.

2.15.2. Injection into the Jugular Vein

The anaesthetised rat is placed on its back with its tail towards the investigator. A small skin incision is made in the neck to one side of midline after hair has been removed with clippers. The large jugular vein is easily seen in young rats but may be overlaid with fat in older animals. In the latter the fat and connective tissue should be cleared by blunt dissection, but the vein itself should not be handled since it will readily constrict. If it does constrict it will regain most of its former size if left untouched for a minute or two. The vein passes under the pectoral muscle where it then bends towards the heart.

For the injection a 25G needle is inserted into the vein at an angle of about 10° through the pectoral muscle, pointing towards the head. Withdrawal of the plunger to draw a little blood into the barrel will show whether the needle lies within the vein. The injection can then be given (Fig. 31). Since the needle passes through the pectoral muscle which acts as a seal, no bleeding ensues when the needle is withdrawn. In addition the pectoral muscle helps to stabilise the needle in the vein.

2.15.3. Injection into the Femoral Vein

The rat, under anaesthesia, is placed on its back with one hind limb firmly held out towards the investigator, e.g. with a rubber band or

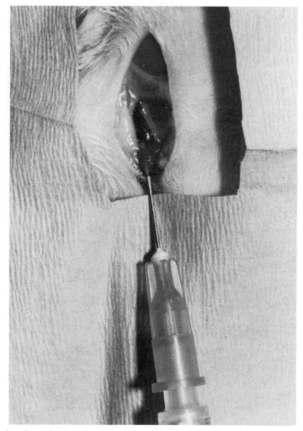

Fig. 31. Injection into the jugular vein. The needle passes first through the pectoral muscle.

string. The hair should be clipped short over the inner surface of the lower leg and the skin swabbed with antiseptic. A small midline skin incision is made in the inside of the leg, and the femoral blood vessels located. The large blue vein lies on the side towards the tail and is covered by fine connective tissue which should be removed over a small area by blunt dissection. Closely accompanying the vein is the femoral artery. The vein is dilated by applying finger pressure over the vein in the region of the groin. The vein is very mobile and the thumb is pressed alongside it while the vein is entered, using a 25G needle (Fig. 32). Alternatively, if an assistant is available to apply the finger pressure, then straddling the vein with two fingers will help steady it (see 2.15).

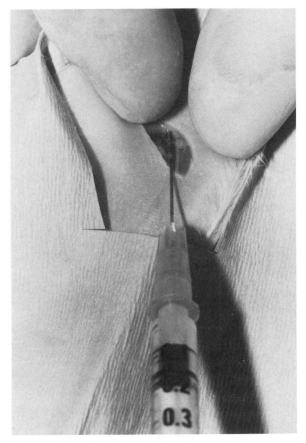

Fig. 32. Femoral vein injection. The thumb and forefinger apply pressure in the groin region to dilate the vein, and tension the skin to immobilise the vein.

The finger pressure is released when the vein is entered and the injection is made. After withdrawal of the needle a cottonwool pledget should be applied to effect haemostasis. The skin incision is then closed with one or two sutures or skin clips.

2.15.4. Injection into the Dorsal Metatarsal Vein

The vein is found on the dorsal surface of the foot. It may be more prominent in one foot than the other. An assistant sits comfortably and holds the conscious rat with one hand while the other extends the leg and holds it steady by the ankle. The leg must not be allowed to move. The dorsal surface of the foot is wet shaved, removing all the hair and swabbed with antiseptic. The assistant puts pressure on the vein at the ankle with his thumb to cause it to dilate. It is then lightly flicked with the fingers to produce further dilation. The toes are now held by the investigator and curved over the lateral tip of the index finger. The skin must be kept tightly stretched to prevent the vein from moving, this is aided by the assistant pulling the skin back at the ankle. The vein is usually easily seen and is entered with a 25G needle kept almost horizontal to its surface, at a point where it just starts to travel up the foot after first crossing the foot and supplying the toes with blood (Fig. 33). The side of the thumb can be used effectively for a steadying support as the needle is pushed into the vein.

2.15.5. Injection into the Sublingual Vein (adapted from Waynforth, 1969)

The two sublingual veins are fairly large and superficial, and are to be found on the lateral aspects of the ventral surface of the tongue. For the injection the rat is anaesthetised in an induction chamber with halothane or methoxyflurane (see 8.2.2) and then removed and laid on its back with its head towards the investigator. The tongue is pulled out with forceps and held at its tip with the thumb and first finger in such a way that it presents a slight convex curvature. The tension on the tongue should not be so great as to cause narrowing of the veins. Since the veins are superficial, they should be entered almost horizontal to their surface (Fig. 34). If the tongue is wet with mucus it should be dried, otherwise the needle will tend to slide on the surface of the

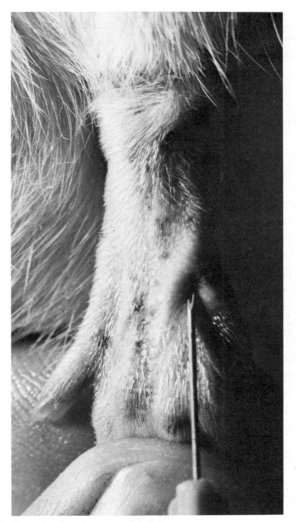

Fig. 33. Injection into the dorsal metatarsal vein.

tongue instead of "biting" into the vein. A 25G needle is the best size to use for rats from weaning age and older, and a 30G needle for smaller rats. After the injection the needle is withdrawn, and the site of the needle entry is covered immediately with the thumb. The thumb is then replaced by a small cottonwool pledget placed on the entry site with forceps. The tongue and the pledget are pushed back into the mouth, the pledget acting as a haemostat. On recovering from the

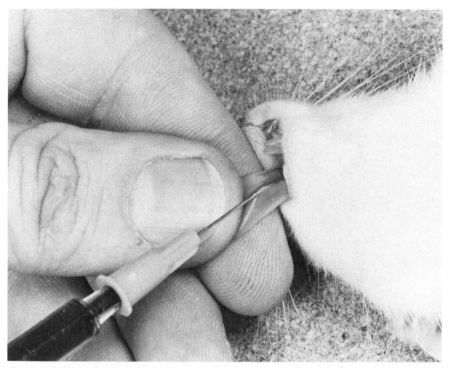

Fig. 34. Sublingual vein injection, note that there are two veins.

anaesthetic the rat spits out the pledget. It should be noted that if the needle does not enter the vein successfully first time it should not be pulled out but manoeuvred within the tongue muscle until it enters the vein. If the needle is withdrawn, bleeding occurs which obscures the site for further injection. If one vein is damaged then the contralateral vein can be used but because of the position of the thumb, entry is made difficult. Haematoma formation is rare and the method can be used for daily or more frequent injections if carried out correctly. The method, with care and practice, is easy to perform but the points made in 2.15 should be taken into account.

2.15.6. Injection into the Penile Vein

This, of course, is applicable only to the male rat! The anaesthetised rat is placed on its back on the table. The glans penis is exposed by sliding the prepuce downwards while pressing at the base of the penis.

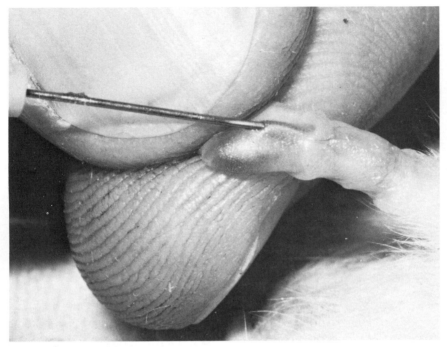

Fig. 35. Penile vein injection.

It is held at the very tip with the thumb and forefinger. If it is not placed under too much tension, the penile vein is easily seen as a distinct central vein (Fig. 35). Entry is made with a 25 or 30G needle. In some cases it may be difficult to tell if the vein has been entered and a small amount of solution will have to be injected to see if it flows freely. If a small bleb is formed this indicates that the needle is outside the vein. After the injection, the needle site is pressed with a cotton wool pledget for a few seconds and the glans is encouraged to retract into the prepuce to prevent further bleeding. In small rats the glans cannot be exposed and cannot be used for injection. This route should only be used under special circumstances, because of the consequences of damage to the vein (see 2.15).

2.15.7. *Intravenous Injection of Neonatal Rats*

It is important that both before and after injection, the neonatal rats and the hands of the investigator should be rubbed over gently with

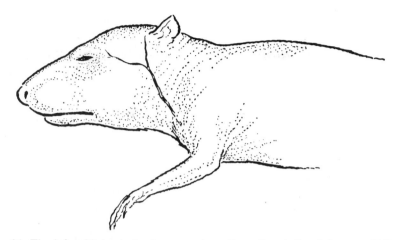

Fig. 36. The infraorbital vein in the neonatal rat (from Campbell and Sargent, 1969, with permission).

the cage bedding material or urine from the mother. This will disguise the smell of the investigator and thereby avoid possible cannibalism of the young rats by their mother. An alternative means is to sedate the mother before removing the young, e.g. with diazepam, 1 mg/kg i.p.

There are three main routes for i.v. injection into neonatal rats and in rats up to about a week old.

2.15.7.1. The infraorbital vein. This runs subcutaneously from the eye to below the ear and then down the neck (Fig. 36). Sighting of the vein is facilitated by wiping the area with glycerol and gently compressing the thorax which increases the central venous pressure and dilates the vein. The vein is entered through the outer lateral edge of the unopened eyelid using a 30G needle. However, the failure rate using manual injection is high because of the ease with which the vein can be completely pierced during discharge from the needle. Greater success can be achieved by using a foot-operated motor-driven continuous flow injection apparatus (or infusion pump). The syringe is placed in the apparatus and connected to the needle via plastic tubing. The operator needs to concentrate only on placing the needle in the vein and holding it steady while the motorised apparatus depresses the syringe plunger.

2.15.7.2. Injection into the femoral vein. This vein is prominent in neonatal rats, and after swabbing the inside of the leg with 70% alcohol the vein can be entered fairly easily through the skin providing the rat is held by an assistant.

2.15.7.3. Injection into the caudal vein. Since those parameters are absent, which make tail vein injection in the adult rat so difficult, viz. tough and scaly or pigmented skin, tail vein injection in the neonatal rat using a 30G needle is much simpler.

2.16. Internal Administration of Substances by the Osmotic Minipump

Implantable osmotic minipumps are available commercially (see Equipment Index). These small devices can deliver solutions continuously for a period of up to 2 weeks without the need for external connection or frequent animal handling. The Alzet minipump system consists of an osmotic pump, a flow moderator and a disposable filling tube (Fig. 37). The minipump is filled with the solution to be administered and then implanted into a prepared s.c. pocket, or it can be placed intraperitoneally. A range of different pumps are now available varying in size from 1.5 × 0.6 cm to 5.1 × 1.4 cm and providing infusion rates of 0.5–1.0 ml/h for 3–14 days. For use, the minipump must be compatible with the chemicals in the solution and this can be tested prior to implantation. The minipump can also be used to deliver solutions intravenously or locally to regions remote from the pump body, such as the brain, by attaching a small catheter and a special flow moderator to the pump. Full instructions for the use of the pump are given by the manufacturers.

2.17. Gastric Infusion (adapted from Waynforth *et al.*, 1977)

This can be carried out in the conscious rat by passing a catheter into the stomach via the mouth using a special muzzle to prevent the rat biting through the catheter. The apparatus is shown in Fig. 38. Item 1 is made from thin-walled needle tubing of o.d. approximately 0.5 mm. The crosspiece fits into the mouth, and the sides run back and downwards on either side of the outside of the jaw to fit into, and be confined by, a plastic snapper clip. Item 2 is fashioned from needle tubing, o.d. approximately 1.4 mm. It passes into and to one side of the mouth for a distance sufficient to prevent the plastic catheter it is to carry from coming into contact with the teeth. It should not be so

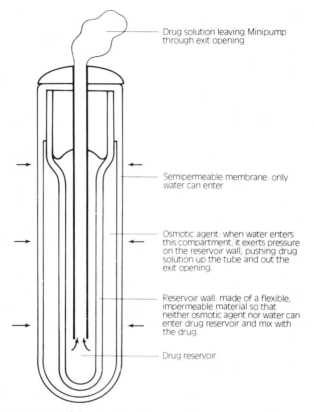

Drug solution leaving Minipump through exit opening

Semipermeable membrane: only water can enter

Osmotic agent: when water enters this compartment, it exerts pressure on the reservoir wall, pushing drug solution up the tube and out the exit opening.

Reservoir wall: made of a flexible, impermeable material so that neither osmotic agent nor water can enter drug reservoir and mix with the drug.

Drug reservoir

Fig. 37. The components of the osmotic minipump. The energy source is a saturated solution which osmotically draws in fluid from the surrounding tissues. This causes compression of the reservoir and ejection of the contained fluid via the outlet.

long that it damages the back of the mouth. The tubing runs back externally along the side of the head, through the opening of the snapper clip, emerging tilted upwards to deflect the catheter away from possible entanglement with the limbs. Item 3 is a lightweight adjustable plastic snapper clip which is available in several sizes. The clip should fit closely around the neck, but not so tightly that it impedes respiration. Item 4 is a polyethylene catheter of a suitable size to pass through the 1.4 mm o.d. tubing. Item 5 is a soldered joint attaching the different size needle tubings. The overall dimensions of the apparatus can be modified for various sizes of rat.

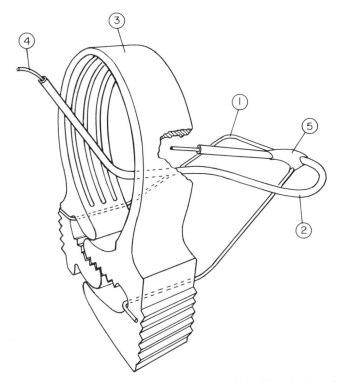

Fig. 38. Apparatus for gastric infusion in the conscious rat (from Waynforth *et al.*, 1977).

Introduction of the catheter into the stomach via the mouth can take place by one of two methods:

(i) Under anaesthesia a needle cannula is placed into the stomach as for intragastric administration (see 2.8.1). The catheter is passed down the cannula which is then withdrawn leaving the catheter in the stomach.

(ii) Again under anaesthesia the catheter is passed into the oesophagus and then into the stomach using the apparatus described for tracheal administration (see 2.11). By-passing the glottis results in the tube passing into the oesophagus.

After the catheter has been placed in the stomach it is threaded through the large diameter mouth tube (Fig. 38, item 2) and then attached to an infusion apparatus. For the duration of the infusion, the conscious animal must be confined in a Bollman cage (see 17.2).

The limitation on the use of the muzzle apparatus is that the rat is unable to eat or drink, and consequently continuous infusion can take place for a maximum of only 24–48 hours. However a liquid food supplement (see 56) can be fed via the indwelling catheter which might make the method more flexible. It is important to appreciate that this technique is likely to cause considerable stress to the animal. In addition to concern for animal welfare, stress may alter the absorption of material infused, and may frustrate the aims of the experiment.

Appropriate modification of the technique described for collecting gastric content (see 23) might also prove a useful means of carrying out continuous infusion into the stomach.

2.18. Cisternal Puncture (and Intracisternal Injection) (adapted from Griffith and Farris, 1942)

The anaesthetised rat is placed over a special wooden stand (Fig. 39) with its head elevated so that the long axis of the body lies at a 50° angle to the horizontal and the four limbs are secured with ties or other means to maintain this position. The head is allowed to hang over the upper part of the stand and is freely moveable. The centre of the stand is hollow to prevent compression of the trachea (N.B. a stereotaxic apparatus can replace the special wooden stand for fixing the rat, see Fig. 26).

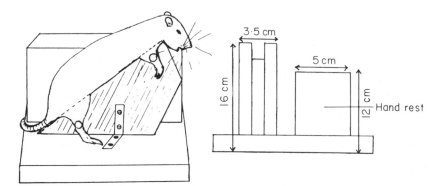

Fig. 39. Wooden stand for performing cisternal puncture and injection.

2.18.1. Cisternal Puncture

The skin is shaved over the posterior aspect of the head. The head is
flexed acutely so that the external occipital protuberance in the neck
region becomes prominent. Directly caudal to this, a depression can be
felt between the protuberance and the spine of the atlas, this locates
the atlanto-occipital membrane. A 24G needle attached to polyethylene
tubing is pushed into the centre of this depression, and when it enters
the cisterna magna a sudden decrease in resistance should be felt. The
cerebrospinal fluid (c.s.f.) flows immediately and usually unhindered
providing the outflow of the polyethylene tubing is kept 3 cm below
the cisternal space. In some cases it may be necessary to withdraw the
needle slightly or provide gentle suction to start the flow. Approximately
0.1–0.15 ml of c.s.f. can be obtained in 3 min. Cerebrospinal fluid is
usually clear, but may be contaminated with a little blood. The degree
of contamination depends on the interval between repeated cisternal
punctures. Repeated punctures with an interval of only a few hours
results in high red blood cell (rbc) contamination. To minimise this an
adequate time interval between cisternal punctures would be 3–7 days.

2.18.2. Intracisternal Injection

The apparatus used for injection into the cisterna is shown in Fig. 40.
A rubber or polyethylene tube fits over end (A) for suction or expulsion
of fluid. Through stopcock (B), either capillary tube (C) or (D) can be
connected. End (E) fits into a 24G hypodermic needle about 1.6 cm in
length. The fluid to be injected is first drawn up until it just enters
tube (C). Stopcock (B) is then turned and tube (D) is filled with the
fluid nearly to the top. The stopcock is then turned to a neutral position
and a cisternal puncture is performed. When the cisterna has been
entered the stopcock is turned to make connection with tube (C). Gentle
suction is employed and c.s.f. mixed with a little of the injection fluid
rises in tube (C). This shows that the needle is correctly placed. The
stopcock is then turned to connect with (D) and the fluid in (D) is
expelled gently to replace approximately the same volume as that of
c.s.f. originally withdrawn.

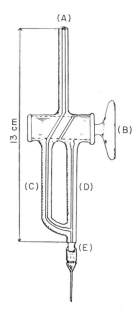

Fig. 40. Glass instrument for carrying out cisternal injection (from Griffith and Farris, 1942).

2.19. Administration of Substances by Inhalation

Carrying out inhalation studies is technically specialised and demanding, and should not be undertaken lightly (WHO, 1978; Phalen, 1984; Lu, 1985; Kennedy, 1989). Some inhalation studies require restraint procedures and environmental conditions which may be stressful, and the welfare of the rat must be considered carefully.

Unlike almost all other routes of administration, it is very difficult to accurately determine dosage via inhalation. For aerosols, the amount of drug deposited in the airway depends on particle size, density of aerosol, airflow, tidal volume, respiration rate and pattern, degree of airway tone, and factors involving airway fluids such as mucus. For example, it should be noted that particles of greater than 9 μm mass median aerodynamic diameter have an extremely small probability of reaching the pulmonary region of the rat. An ideal particle size is below 3 μm, but very small particles will not be absorbed and will be expired. For gases and vapours, all the physiological factors will play a role, and the limiting factor may be the absorption of the drug into the airway fluids. In real terms, the only indication of the dose delivered to the rat is the amount that can be measured in the blood and tissues. Blood sampling is thus an integral part of many studies. It is therefore essential to standardise dosing regimes and to calibrate equipment such as nebulisers carefully so as to reduce variables to a minimum. It is very important to perform calibrations with all associated tubing in place, since any change in air resistance in the system will tend to alter output.

Aerosols may be generated in many ways. If possible, it is preferable to use soluble compounds since generation of powder aerosols is technically difficult. Ultrasonic or pressure-driven jet-type nebulisers may be used (see Equipment Index). When nebulising drug solutions one should be aware that the physical effects of nebulisation may alter the apparent solubility of drugs. For example, passing a solution through a jet nebuliser may drop the temperature of the solution by up to 20°C, and this may cause drugs to fall out of solution. This effect is not so apparent with ultrasonic instruments, but evaporation of solvent from the aerosol phase in either type may result in small particles of solute falling back into the drug solution reservoir. This will cause concentration of the remaining solution, or may induce precipitation. Both these effects will alter the amount of drug delivered to the animal. It is essential to test your drugs for this effect, and to use fresh solutions for each nebulisation whenever possible. Some examples of methods which can be used to administer aerosolised drugs to individual or small numbers of rats are as follows.

2.19.1. Administration to Anaesthetised Rats

The trachea is exposed via a midline incision in the ventral skin of the neck and a small lateral incision made in it. The tracheal incision should be only large enough to allow the insertion of a polyethylene or similar tube (e.g. infant nasogastric feeding tube, size 8 French is suitable) for a distance of 5–10 mm. The length of the tube should be as small as possible to reduce dead space to a minimum and thus reduce rebreathing of respired air. The tracheal cannula is provided with a side arm constructed from a simple nylon T-piece. A bias airflow is generated across this T-piece, provided by a small vacuum pump of the piston or oscillatory type. This draws air through the system, thus enabling the animal to respire from a passing aerosol flow generated by a nebuliser. The airflow generated need only just exceed the peak airflow during inspiration. In some experiments it is necessary to attach pneumotacho-graphs to the trachea, but these may have an internal volume greater than the tidal volume of a rat, causing problems with dead space and subsequent rebreathing of air. However, the bias flow system shown (Fig. 41) eliminates any problem with this dead space since the volume is continually purged. Depending on the type of pump used, pressure pulses may be produced which would interfere with the pneumotacho-graph signal. This problem can be minimised by using small bore tubing to increase impedance between the trachea and the pump, and incorporating a reservoir and a pinhole to damp out these pulses. The pinhole can be conveniently produced from a piece of hard plastic tube, often found as a seal on the end of male Luer connectors or three-way taps, which is pierced with a 25G needle. It should be noted that the dimensions given in Fig. 41 are from a working system, but these may need to be adjusted to suit individual needs. Fitting a pneumotachograph is useful since tidal volume and respiration rate can be determined from it, and dosing can be controlled according to respired volume and/or number of breaths.

Two types of pneumotachograph are commonly in use, the screen type, and the Fleisch type which is electrically heated. The Fleisch type (Fig. 41) is more suitable for aerosol dosing, since the heated element prevents build-up of moisture in the pneumotachograph. Screen type pneumotachographs of a size suitable for use with rats tend to block when aerosols are passed through them. If a more accurate technique is required then it is better to use a piston-type respiratory pump to deliver a number of breaths of known volume. This can be done either via a tracheal cannula, or by inserting a tightly fitting cannula into one nostril while blocking the other with wax and sealing the mouth with tape. In this technique the nebulisation chamber must be incorporated into the output tube of the respiration pump, making the technique only suitable for ultrasonic nebulisers. This technique has the advantage of standardising the volume delivered, but it must be restated that the system must be calibrated with all its associated tubing at the required respiration rate and tidal volume to estimate the dose given. It should also be noted that some of the dose will not be absorbed, and will be exhaled. Therefore blood levels will need to be measured in samples taken via a catheter placed in, for example, the femoral vein, to determine the amount of drug actually absorbed by the body. This technique can also be adapted for artificially respired animals, but again

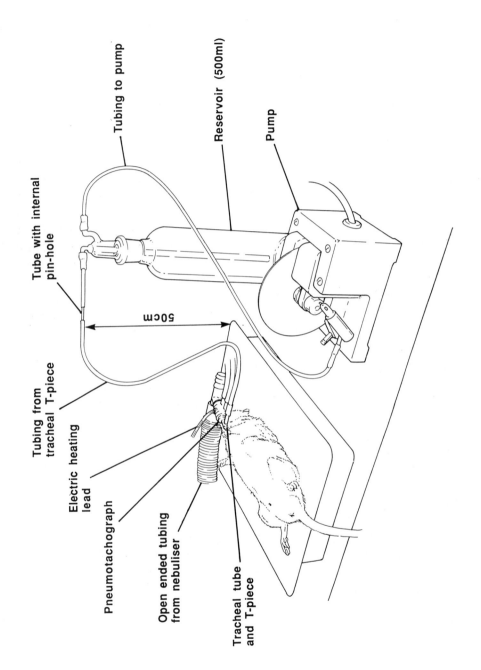

Tubing to pump

Reservoir (500ml)

Pump

Tube with internal pin-hole

50cm

Tubing from tracheal T-piece

Electric heating lead

Pneumotachograph

Open ended tubing from nebuliser

Tracheal tube and T-piece

care must be taken over the design of the system so that drug delivery stays constant.

2.19.2. Administration to Conscious Animals

Controlling the dose in conscious animals is even more difficult than in anaesthetised ones. The simplest method is to place rats in a chamber such as an 8 l plastic desiccator with one or two 8 mm holes drilled in the base to allow for the escape of aerosol. Up to four adult rats may be placed in a chamber of this size. The output tubing of the nebuliser is connected directly to the chamber and aerosolised drug passed in for a known time. Signs of distress during dosing can easily be observed using this system and administration discontinued. For more direct delivery of drug, the rat is placed in a restrainer with its head within a cone. The cone contains several holes to allow free escape of aerosol. The output tube of the nebuliser is attached directly to the cone and aerosol passed over the animal's head while it is breathing normally. This technique is shown in operation in Fig. 42. However, additionally in this picture an arrangement is shown in which pairs of animals are placed in restrainers on a tilt table in order to study the effect of postural change on cardiovascular parameters during drug delivery. Aerosols are passed over the animal's nose and head via the block cone. If required, respiration rate can be determined by connecting a low pressure transducer to one of the holes in the cone and analysing the pressure output. Blood samples are taken via a previously implanted venous catheter to estimate drug "delivery" to the animals. It is important in these studies to provide suitable extraction of exhaled and waste aerosols, by enclosing the restrainers and nebuliser in a suitable fume cupboard or exhaust enclosure. In large-scale studies, such as for toxicology, a plenum chamber fitted with many restraining cages configured in various ways has been used and allows long-term multiple exposures to inhaled drugs (see earlier references).

It should be noted that exposure of rats to inhaled drugs for prolonged periods may cause irritation or damage to the respiratory tract, or the

---

Fig. 41. Arrangement for inhalation dosing of anaesthetised rats. A nebuliser (not shown) aerosolises the drug which is sucked through the open-ended tubing by the action of the pump. The rat breathes in the aerosol via the electrically heated pneumotachograph.

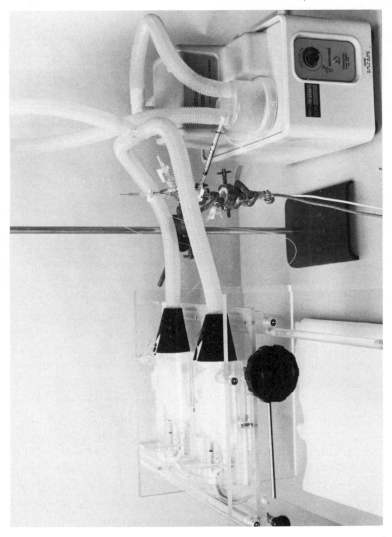

Fig. 42. Arrangement for inhalation dosing of conscious rats. The nebuliser is connected via tubing to the black cone of the special restrainers (see Equipment Index), allowing the aerosolised drug to flow over the animal's head. Probes placed in the black cone and connected to a transducer (centre) measure respiration rate and volume.

drug itself may be inherently irritating or toxic. This may cause an increased susceptibility to respiratory infection and it is advisable to keep such animals in a barriered room or, for example, in a flexible film isolator between exposure periods.

Chapter Two

METHODS OF OBTAINING
BODY FLUIDS

3. BLOOD

3.1. Quantitative Aspects of Blood Collection

An adult rat has a blood volume of approximately 70 ml/kg, and it is possible to withdraw up to 10% of this volume at any one time without causing significant effects. Removal of larger volumes can result in demonstrable stressful effects, for example elevation of corticosterone concentration after removal of 10–25% of circulating volume (Wiersma and Kastelijn, 1985). Removal of more than 20–25% of blood volume usually produces signs of hypovolaemia, and can precipitate cardiovascular failure ("shock"). If the rat survives the removal of excessive volumes of blood, it is unlikely to be of value as an animal model. For this reason, as small a volume as is consistent with the aims of the experiment should be removed. Even when relatively small volumes of blood are withdrawn, although the blood volume is restored within 24 hours, some constituents (e.g. erythrocytes and reticulocytes) may not have returned to normal until after about 2 weeks (Schermer, 1967). It would therefore seem sensible not only to minimise the volume of blood withdrawn, but in some circumstances to refrain from removal of further quantities for 2 weeks. When frequent blood sampling, or removal of more than 10% of blood volume, is necessary as part of the study, the potential adverse effects of this can be minimised by replacing the blood removed by transfusion. Blood collected from another rat of the same strain can either be used immediately or collected in acid citrate phosphate dextrose solution (ACD, see 59) and processed and stored for use a day or more later (Ellis and Desjardins, 1982, Wiersma and Kastelijn, 1985). For example, use of a series of intermittent transfusions would allow the collection of small quantities of blood for biochemical analysis (say 0.2 ml) every 5 min over an 8–12 hour period. Removal of up to 1% of blood volume daily for a short time is unlikely to have any haematological consequences and can therefore be carried out regularly if required.

The site from which blood is withdrawn, the anaesthetic used and the effects of any stress induced are important aspects for consideration if a study of haematological parameters is intended. Failure to consider these or to standardise methodology will lead to considerable variability of the results. The effects of stress are well documented (see Kraus, 1980). For best results under these circumstances, Archer and Riley (1981) recommend the use of halothane, if anaesthesia is required, and the use of the jugular vein or ophthalmic venous plexus for the removal of blood.

The site of blood sampling may be of considerable importance for obtaining accurate estimates of the concentration of drugs and implications in pharmacokinetics, pharmacodynamics and toxicology have been discussed by Chiou (1989).

3.2. Methods not Requiring Anaesthesia of the Rat

3.2.1. Obtaining Blood by Decapitation

Decapitation of rats is a procedure which many research workers find unpleasant and distressing, but when carried out competently it is an extremely rapid and humane method of killing the animal and obtaining a moderately large quantity of blood.

Two methods of decapitation are commonly used. The rat can be first stunned with a blow to the back of the head (see 47) and then the neck severed either completely or partially using strong scissors. Alternatively, the head can be removed by use of a guillotine (see 47). It is important that the operator is experienced in methods of handling and restraint. If expert assistance is not available, then the rat may be placed in a polythene restraining device (Decapicone—see Fig. 8 and Equipment Index) prior to use of the guillotine. It is necessary to place the severed neck very quickly over a suitable collecting vessel because of the copious and strong flow of blood which ensues. The blood will be contaminated with tissue fluid and the yield may not be good.

3.2.2. Bleeding from the Tail

A frequently used and relatively simple method of obtaining blood samples from the tail of newly weaned to adult rats is to insert a needle into a lateral vein. To carry out the technique the rat is warmed in a "hot box" for about 15 min (see 2.15). The rat is then removed from the hot box and either placed in a restraining device or restrained manually by an assistant who places it on a waist-high bench with its tail draped over the edge. The assistant gently twists the tail to expose a lateral vein and puts pressure on the base of the tail to act as a tourniquet. The tail veins are clearly visible in young rats, but in older animals they are more difficult to locate, although cleaning with a damp swab will often be found helpful (also see p. 44). If the technique is being carried out without an assistant, the warmed animal is placed in a restraint apparatus and a rubber band or piece of string (Fig. 28) (Minasian, 1980) is applied to the base of the tail as a tourniquet. The tourniquet should be released to allow the blood to flow freely once the needle is in the vein. A 23 or 21G needle and 1 ml syringe can be used for collection of blood, but a particularly useful technique is to use a 21G × 1" butterfly needle which is modified by removing all but 5 mm of its plastic cannula. The needle is then placed in the vein at a shallow angle, parallel to it at the point where the operator is holding the tail with his left hand, about two thirds of the distance down the tail (Fig. 43). Successful entry in the vein is accompanied by a flow of blood. During collection of the blood, the cut end of the cannula should touch the collecting vessel as this seems to promote blood flow. Removal of the needle and light finger pressure will stop further bleeding. Further samples of blood can be obtained by repeating the procedure while using a slightly more anterior portion of the vein, or using the other lateral vein. It should be noted that variations in obtaining samples may occur and an animal may require a longer period of heating. If this is the case the rat must be observed for signs of heat stress, manifested

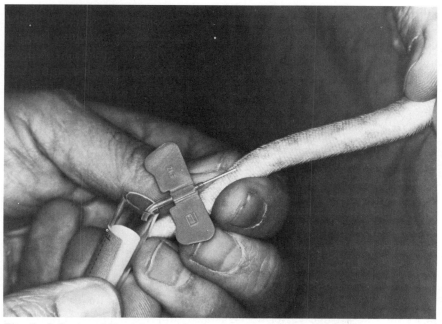

Fig. 43. Collection of blood from a tail vein with a Butterfly needle (by kind permission of D. Gask).

by salivation or collapse. Since with experience, each animal can be bled within 2 or 3 minutes, placing animals in the hot box sequentially will allow a large batch of animals to be bled with minimum delay.

An alternative technique for obtaining small (<0.5 ml) blood samples involves complete transection of the tail, approximately 5 mm anterior to its tip. Blood flow will be increased if the tail is first warmed for 1–2 min in hand-hot water (40°C), and also if the tail is "milked" by sliding the fingers down the tail from its base during bleeding. It should be noted however that "milking" the tail will set up an inflammatory leucocytosis thereby producing an abnormal increase in the number of white blood cells in the blood that is obtained. Even greater volumes of blood (at least 1.5 ml) are obtained if the animal is placed in a 37–40°C oven for 15 min. Bleeding usually ceases of its own accord, but slight finger pressure will also be effective. However, a styptic pencil (silver nitrate) may be needed to stop further haemorrhage, if whole-body heating has been employed. Alternatively, a piece of collagen haemostat ("Xenocol"—see Equipment Index) can be pressed on to the tail tip. More anterior portions of the tail can be cut if blood is required on subsequent occasions, but only two or three samples should be obtained by this method since repeated cutting of the tail tip will result in trauma to cartilage and eventually to the coccygeal vertebrae, a procedure which is both too painful to carry out in a conscious rat, and one which results in unnecessary injury to the animal. Alternatively, for closely spaced serial samples, the original incision wound can be opened up by repeatedly removing the clot that forms.

A less emotive and possibly less painful alternative for the animal is to incise the vein using a scalpel blade. The tail is held on the table surface or in a groove prepared in a rubber bung. The tail need not be warmed if only a small quantity of blood is required but it may need to be milked for the flow to start. The blood is collected in a capillary tube or in a Microvette (Fig. 1) (see Equipment Index). This is a small plastic test tube with a pointed end which allows blood to be collected by capillary action. It has a volume of 300 μl. When the blood has been collected the tip is sealed with the small stopper which is supplied and the open end with the integral cap. The closed tube is then centrifuged in its own centrifuge vessel which is also supplied (Fig. 1). The serum sample is pipetted off with a micropipette. This method has been used successfully to collect blood from sentinel rats in a disease monitoring programme, even within the confined space of a flexible film isolator.

Tail blood can also be obtained by using a vacuum-assisted method (see Nerenberg and Zedler, 1975). To prepare the apparatus required (see Fig. 44), a 250 ml Liebig condenser jacket is cut at the blind end to form two open ends with a distance of 14 cm between them. The upper end is covered with a soft rubber grommet (2.8 cm o.d.). A test tube (25 × 75 mm) is introduced through the lower (threaded) opening, following which the threaded cap is tightened over the contained rubber grommet (2 cm o.d.) to form an air-tight seal with the test tube. The side arm is connected to a vacuum line (35 cm of water, negative pressure) via plastic tubing containing a T-connector.

The rat is first placed in a restrainer with its tail protruding. A vein in the distal end of the tail about 1 cm from the tip is incised longitudinally for about 1 cm in length, after first smearing the incision area with a little petroleum jelly to aid droplet formation of the blood. The test tube and cap are removed from the condenser jacket, and the jacket is slipped over the tail to fit snugly against the buttocks. The tail should protrude for about 2 or 3 cm from the end of the jacket. The tube is reinserted over the tail and the cap is tightened. The suction is

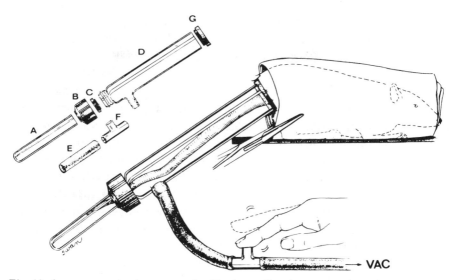

Fig. 44. A vacuum assisted apparatus for blood sampling from the tail: (A) test tube, (B) threaded cap, (C), rubber grommet, (D) modified Leibig condenser jacket, (E) plastic chamber connecting bleeding apparatus to vacuum line, (F) T-connector, (G) rubber grommet (from Nerenberg and Zedler, 1975).

started and an air-tight seal is produced if the jacket is momentarily pushed against the buttocks. The intensity of the vacuum and the rate of bleeding is controlled by intermittent finger occlusion of the open end of the plastic T-piece.

A volume of blood of 0.5–1 ml can be collected quickly, and at least eight serial samples of blood can be collected within 24 hours by opening up the original incision wound. Since the apparatus is not contaminated with blood, a number of small samples of blood, with or without the addition of heparin or other additives, can be collected at one time simply by changing the collection tube.

3.2.3. Bleeding from the Jugular Vein

With practice a conscious rat can be bled easily from the right jugular vein. The animal is picked up by the scruff of the neck and held ventral side up against the investigator's body to prevent it struggling. The head points towards the investigator. The neck is kept stretched out and the hair over the right side of the neck is removed with clippers. Using the line of the jaw as a guide and with the jaw placed against the hub of a short No. 21G × ⅝″ needle attached to a 2 ml syringe, the needle, placed at a very shallow angle, is slowly pushed through the skin at the angle made by the neck with the shoulder (Fig. 45). While the needle is being pushed in the syringe plunger should be slightly withdrawn until a show of blood in the needle hub or syringe point indicates puncture of the vein. The blood sample is then withdrawn slowly—or an injection can be made if desired. Usually no bleeding or haematoma ensues when the needle is removed. Although serial samples can be taken, the interval between samples should not be less than 24 hours and preferably 48 hours. In experienced hands bleeding from the jugular vein in the conscious rat may be less stressful than by some other methods (e.g. tail vein or orbital sinus) as determined by the lower level of circulating plasma corticosterone found for the former.

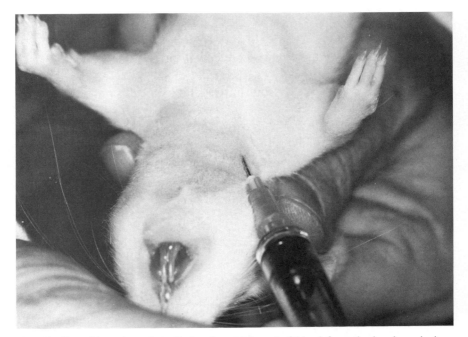

Fig. 45. Site of insertion of needle for the withdrawal of blood from the jugular vein in a conscious rat (by kind permission of D. Gask).

3.3. Methods Requiring Anaesthesia of the Rat

3.3.1. Cardiac (Heart) Puncture

Cardiac puncture is best used only for terminal blood sampling because of the risk of damage to the myocardium. As an alternative, reasonably large volumes of blood (1–1.5 ml) can readily be obtained from the tail vein (see above). If the rat is intended to recover from the cardiac puncture, a 23 gauge needle should be used to minimise the damage to the myocardium. For terminal blood sampling, a larger gauge needle (21G) can be used. To carry out cardiac puncture, the rat must be anaesthetised and placed on its back with its head positioned to the left of the investigator. The thumb and fingers of the left hand are placed on either side of the thorax which is then slightly compressed. The hypodermic needle attached to a syringe is inserted under the xiphoid cartilage which is raised slightly with the index finger of the left hand (Fig. 46). One of four procedures can now be employed for entering the heart:

(i) Immediately the needle is pushed under the skin the plunger is pulled back slightly to create a vacuum. Holding the syringe at an angle of about 30° to the horizontal, it is pushed slowly forwards until a small show of blood in the syringe indicates that the heart has been entered.

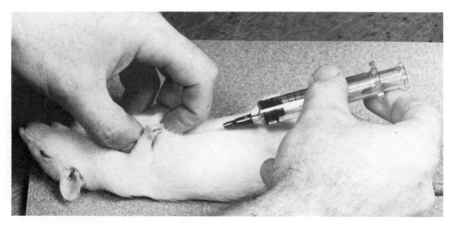

Fig. 46. Cardiac puncture; note that the needle is pushed under the xiphoid cartilage which is raised by the index finger of the left hand.

Occasionally the force of the blood will automatically move the plunger backwards, but it is more usually necessary to aspirate the blood. To do this the plunger can either be withdrawn by holding it with the thumb and middle finger and pushing against the flange of the barrel with the index finger, or by sliding the left hand up to hold the barrel while the plunger is withdrawn with the other hand (Fig. 47).

(ii) After entering the skin the syringe is held lightly around the top of the barrel and moved forwards under the top of the rib cage, almost horizontal to the body surface. The position of the top of the heart can then be gauged when the needle begins to transmit the heart beat to the hand holding the syringe. At this juncture the syringe is raised to about 30° and pushed into the heart by only 3 or 4 mm.

(iii) The syringe, at an angle of about 30°, is slowly pushed into the heart, entry often being signalled by a momentary easing of the effort required to push the needle through the heart muscle. The plunger should be withdrawn once or twice during the forward movement to ascertain when the heart has been entered. This last method is used only by experienced personnel and the absolute novice is advised to start with method (ii), possibly combining it with method (i). Even the experienced investigator may have to resort to method (i) or (ii) on occasions when success by one of the other methods eludes him.

(iv) As an alternative, a "Vacutainer" blood sampling system (Fig. 1, see Equipment Index) can be used. The procedure described in (i) above is followed; after inserting the needle through the skin and into the

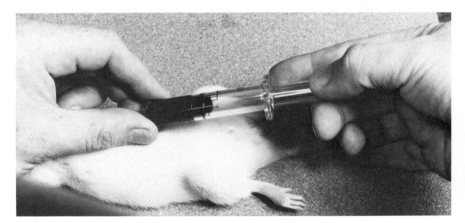

Fig. 47. Cardiac puncture; note how the syringe is held while withdrawing blood.

thoracic cavity, the sample tube is pushed on to the needle mount, creating vacuum suction. The needle is then advanced until it pierces the heart, when blood will be sucked into the sample tube.

Two alternative approaches to cardiac puncture can also be employed. In the first of these, the rat is laid on its right side. The heartbeat can be palpated with the finger and thumb placed on either side of the chest, just posterior to the rat's elbow. If the animal is to be allowed to recover, a 2 ml syringe and 23G needle will be found to be most suitable. The needle should be introduced into the left side of the chest, perpendicular to the chest wall, immediately over the area in which the heartbeat is most easily palpable (Fig. 48). As the needle is introduced, gentle suction should be applied with the syringe and the needle advanced slowly until blood is obtained. Once in position, it is important to avoid moving the needle tip whilst continuing to aspirate blood.

In the second approach, the rat is laid on its back and the thumb and fingers are placed on either side of the chest while gently compressing it. A 23G needle attached to a syringe is pushed vertically through the

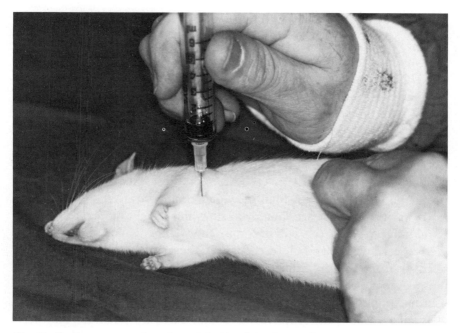

Fig. 48. Cardiac puncture through the left side of the chest wall.

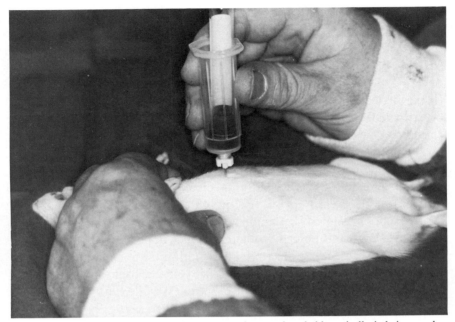

Fig. 49. Cardiac puncture through the sternum. A vacutainer held vertically is being used.

sternum at about the middle while slowly withdrawing the plunger (Fig. 49). Entry into the heart is indicated by a show of blood.

In any of the above methods, if blood cannot be withdrawn, then the heart may either have been pierced completely, in which case slow withdrawal of the syringe will bring the needle back into the heart cavity, or it may have been bypassed, in which case a fresh attempt should be made. No more than two or three fresh attempts should be made to enter the heart as each attempt causes some haemorrhaging from the heart, the sum total of which could prove fatal. Withdrawal of blood from the heart should be slow and steady, at a rate of about 2 ml/min or less, as this will maximise the volume that can be withdrawn before cardiovascular failure occurs. If the animal is to be allowed to recover from the procedure, the volume of blood removed should be limited to 10–15% of circulating volume, approximately 7–10 ml/kg. Cardiac puncture can be carried out at weekly intervals if required, but a smaller (even daily) interval, although certainly possible, may have unpredictable and deleterious consequences.

3.3.2. Bleeding from the Jugular Vein

The anaesthetised rat is placed on its back with its tail towards the investigator. The jugular vein is exposed and entered through the pectoral muscle, the needle pointing towards the head (Fig. 31, see 2.15.2). Blood is withdrawn slowly. After removal of the needle the skin incision is closed with one or two skin clips or sutures. Repeated sampling can be carried out by using both jugular veins alternately.

3.3.3. Bleeding from the Posterior Vena Cava

The anaesthetised rat is laid on its back, and a V-cut is made through the skin and abdominal wall starting at the base of the abdomen and proceeding diagonally across on each side to end up at the dorsolateral edges of the thorax. The flap of skin is moved onto the chest wall and the entire gut is reflected over to the left of the rat (right of the investigator). At the level of the kidneys the widest part of the posterior vena cava will be found, and the liver is pushed forward so that a good view is obtained. The vena cava is entered at this level using a 19 or 21G needle. To avoid the vessel moving forward upon entry, it should be straddled with the first and second fingers of the free hand (Fig. 50). Care should be taken not to pierce the entire vessel. Blood should be withdrawn very slowly until the vessel wall collapses. Withdrawal should be halted temporarily while the vessel refills with blood, and withdrawal can then continue. This sequence of events can occur three or four times until no more useful amount of blood can be withdrawn. This method can only be used for terminal exsanguination of rats.

3.3.4. Bleeding from the Dorsal Aorta

An alternative technique for obtaining large volumes of blood when the animal is not to be allowed to recover is to withdraw blood from the dorsal aorta. The approach to the collection of blood from this vessel is identical to that from the posterior vena cava except that the aorta is entered just anterior to its distal bifurcation into the common iliac arteries.

A second method of collecting blood from the aorta entails blunt dissection of the aorta for a short distance anterior to its bifurcation

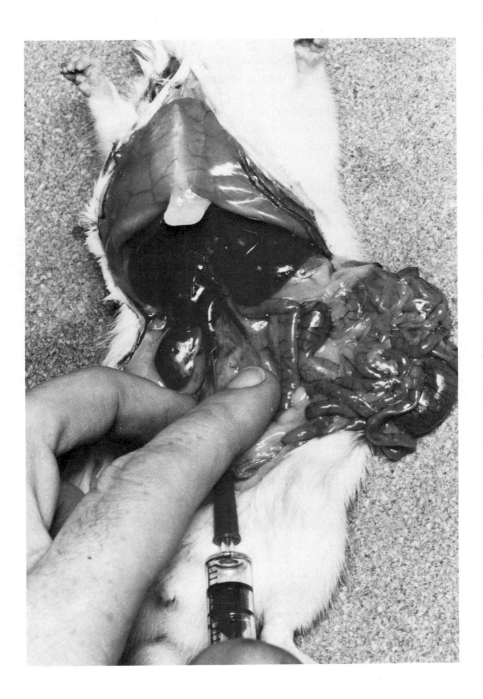

and then clamping the aorta at this point. The front of the rat is then elevated slightly and the aorta is completely transected at its point of bifurcation; the cut end is placed into the mouth of a collection vessel or tube and the clamp is then released slowly to avoid a too vigorous spurt of blood. Several millilitres of blood can be collected quickly. Collection from the aorta (either method) is a terminal procedure.

3.3.5. Bleeding from the Axillary Vessels

An alternative method of terminal blood sampling, which requires virtually no technical skill, is to collect blood as it pools after severing the axillary vessels. The anaesthetised rat is laid on its back with its tail towards the investigator. One forelimb is maintained stretched out with a pin through its foot. An incision is made in the skin at the side of the thorax proceeding into the angle of the forelimb (axilla region). The bottom skinflap is held with forceps to form a "cupped" area. With scissors, all blood vessels in the angle of the forelimb which may include the jugular vein are cut quickly. This can be done "blind" by cutting deep into the axillary region. Blood wells up, accumulating in the cupped area, and can be collected with a Pasteur pipette (Fig. 51). A large volume of blood is obtained quickly, but there will be contamination with tissue fluids.

3.3.6. Bleeding from the Ophthalmic Venous Plexus (Orbital Sinus)

Rats should be anaesthetised. If an inhalation anaesthetic is used in an induction chamber, the rat is removed and will remain anaesthetised for a sufficient period of time for the technique to be completed, providing it is done quickly. Some investigators may be averse to using this technique because of its emotive nature. Also contrary to what was first thought, it has been shown that some damage to organs and tissues associated with the back of the eye may occur (McGee and Maronpot, 1979). The technique should only be used if there is no suitable alternative such as bleeding from the tail vein (see 3.2.2).

Fig. 50. Bleeding from the posterior vena cava; the vein is straddled by the operator's fingers to prevent it from moving forward during puncture with the needle.

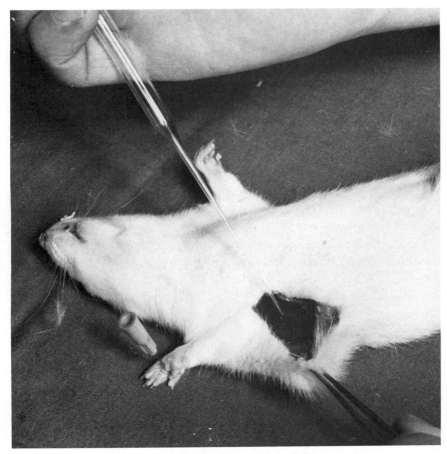

Fig. 51. Collecting blood after severing blood vessels in the axilla.

The anaesthetised rat is held as in Fig. 52 with its head pressed very firmly against a table surface. No movement of the head must be allowed. By pressing down with the thumb and forefinger just behind the eye and pulling back the skin, the eyeball is made to protrude. The thumb should be placed so that it is also occluding the jugular vein. A fine-walled Pasteur pipette, either obtained commercially or prepared by drawing out glass tubing to a fine capillary 2.5–5 cm long with an o.d. of 1–2 mm for large rats and 1 mm or less for small animals, is positioned at the inner corner of the eye, beside the eyeball (Fig. 53). The pipette is then slid a few millimetres forward gently but firmly

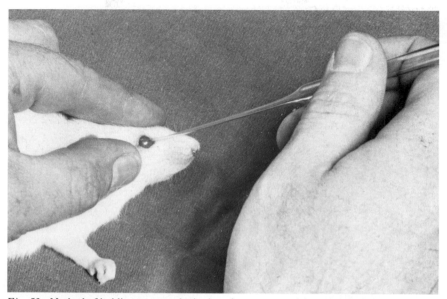

Fig. 52. Method of holding an anaesthetised rat for puncture of the ophthalmic venous plexus; here the pipette is positioned for withdrawal of blood.

along the side of the orbit to the ophthalmic venous plexus (orbital sinus) which lies at the back of the orbit (Fig. 54). The blood vessels of the orbital sinus are extremely fragile and rupture on contact with the pipette tip, allowing blood to enter by capillary action. It may be necessary to rotate the pipette during its forward passage in the orbit to encourage the blood to flow. If no blood is obtained, the pipette should be withdrawn very slightly and rotated at the same time. At least 1 ml of blood can be obtained easily and no suction is required. An alternative, which will allow the collection of more blood, is to insert a 100 μl capillary tube into the eye with the head placed on the edge of the table. Once blood starts to flow the head can be tilted so that the capillary tube is facing downwards and blood can be allowed to drip into a collecting vessel. After bleeding, the pipette or tube is removed and the pressure on the eyeball is released. Bleeding usually stops immediately, particularly if the eyelids are closed and any surplus blood is removed with gauze. Using each eye alternately, two bleedings a week can be carried out. More frequent bleedings should not be attempted.

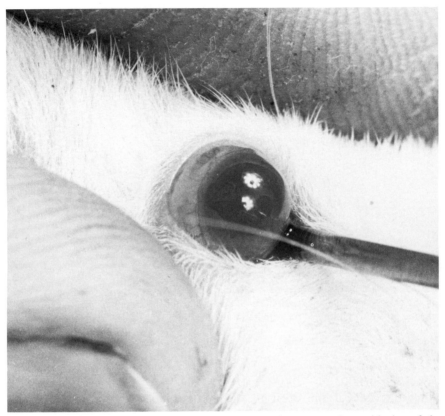

Fig. 53. Collecting blood from the ophthalmic venous plexus, note the positioning of the pipette at the inner angle of the eye.

3.4. Collection of Blood by Means of a Chronic Indwelling Catheter (adapted from Scharschmidt and Berk, 1973)

The methods for long term catheterisation of blood vessels both for the purposes of injection and for the withdrawal of blood are described in Section 22. However, a method will be described here which permits the rapid and frequent short-term sampling of arterial blood using an easily assembled device which can be made from readily available and inexpensive materials. Materials needed are (1) a plastic three-way stopcock, (2) a 23G scalp vein infusion set (Butterfly needle), (3) a 19G

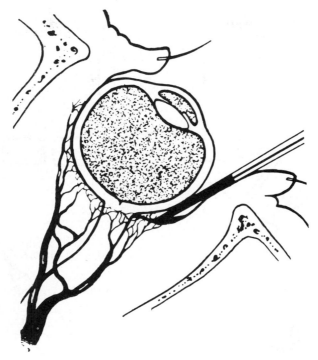

Fig. 54. Diagram of the ophthalmic venous plexus and the positioning of the puncturing needle or pipette (from Riley, 1960).

needle, (4) a 5 ml syringe, (5) heparinised glass capillary tubes, (6) polyethylene tubing of the following sizes:

	i.d. (mm)	o.d. (mm)	length (mm)
a.	3.17	6.35	12.7
b.	1.59	3.17	4.8
c.	2.38	3.97	4.8
d.	0.79	2.38	5.1
e.	0.58	0.96	5.1–10.2

The device is illustrated in Fig. 55, and the steps in the modification of the individual components are shown in Fig. 56.

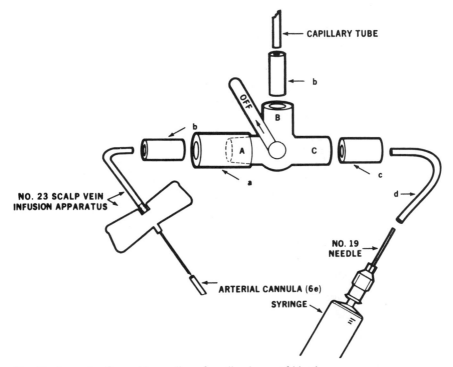

Fig. 55. Apparatus for rapid sampling of small volumes of blood.

For use, the stopcock handle is turned to position B and the system is flushed with saline. The stopcock is now turned to position A and the apparatus is ready for arterial catheterisation with the infusion set. Either the dorsal aorta (see 22.5) or the carotid artery (see 22.3) can be most conveniently catheterised.

The length of the catheter should not be longer than 8 cm which will have a dead space of about 100 µl. After catheterisation the plastic butterfly of the infusion set is taped to the animal board to remove the risk of accidental withdrawal of the catheter.

When the sample is to be taken the stopcock is turned to position B, opening the connection between A and C, and the syringe plunger is withdrawn until undiluted arterial blood appears distal to stub C. A heparinised glass capillary (o.d. 1.6 mm, length approximately 75 mm) is inserted into site B and the stopcock handle is turned towards position C until the capillary tube is filled (2–3 s). When the tube is filled the

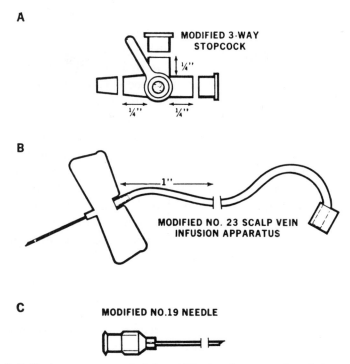

Fig. 56. Individual components and their modification for preparation of the apparatus in Fig. 55 (from Scharschmidt and Berk, 1973).

stopcock is turned to position B and the system flushed with saline until the polyethylene tubing is clear. The stopcock handle is then repositioned at A until the next sample is to be taken. Capillary tube samples can be obtained at intervals of 30 s or less.

3.5. Collection of Blood from Neonatal Rats (adapted from Gupta, 1973)

This can be done conveniently by cardiac puncture, and the following pieces of apparatus are required: a 5 ml syringe barrel, a double-ended collecting hypodermic (Vacutainer) needle (see Fig. 1); a rubber stopper from a Vacutainer tube; a collecting tube to fit inside the syringe barrel; a shirt button with four holes or a similarly perforated piece of metal.

The shirt button is placed inside the syringe barrel and lies at the bottom to provide a flat base without interference of the vacuum that is to be supplied subsequently. The tube is placed in the barrel on top of the button. The rubber stopper is inserted into the open end of the syringe barrel. The short end of the needle is pushed through the stopper into the tube and must protrude into the tube for no more than 2–3 mm. The nozzle of the syringe barrel is connected via tubing to a vacuum line. A vacuum of less than 5 cm of mercury is required.

The rat is held by its back skin, on its back, with its head hanging down and towards the investigator. With the vacuum on, the needle is inserted into the thoracic cavity, through the thoracic inlet, and passed forward very slowly until a trace of blood is seen in the collection tube (Fig. 57). The needle is then held steady until between 0.2 and 0.7 ml of blood is collected depending on the age of the rat. Rats up to 16 days of age can be bled conveniently by this method.

4. COLLECTION OF SEMEN (adapted from Birnbaum and Hall, 1961)

Semen containing viable spermatozoa is obtained by electroejaculation. Under natural circumstances rat semen clots immediately following ejaculation due to the coagulating gland enzyme vesiculase. During mating the finding of such a clot in the vagina of the female rat (vaginal plug) is an indication that mating has occurred (see 55). To prevent clotting, electroejaculation can be carried out in rats in which the coagulating glands have been removed, but the necessity of doing this can be avoided, at least in the Wistar rat, by using carefully controlled conditions, and sperm, free of coagulum, can be collected.

The components required for use in rats 4–24 months of age are: (i) a single bipolar rectal electrode which should be about 4.8 mm in diameter with the tip and ring electrode contacts 2.5 cm apart; (ii) an audio oscillator with the ability to supply sine waves of frequency about 10 to greater than 60 cycles/s and any voltage up to 20 V into a load of 600 Ω or more.

To carry out the procedure the conscious rat is tied, ventral side down, onto an animal board by its four limbs. The tail and hindquarters

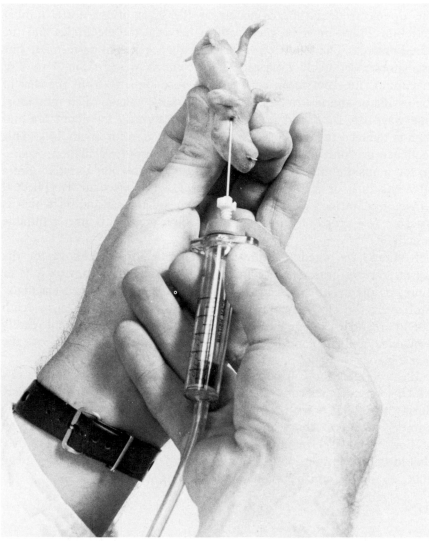

Fig. 57. Vacuum-assisted apparatus for cardiac puncture in the neonatal rat (adapted from Gupta, 1973).

are raised and the rectal probe, lubricated with a water soluble lubricant (e.g. KY jelly—see Equipment Index) to which is added a small amount of salt to increase its conductivity, is inserted slowly into the rectum until the rear ring electrode is just inside the rectum. This ensures that the tip electrode is approximately at the level of the forth lumbar vertebra. There must be good contact between the electrodes and the rectal mucosa. The following electrical sequence is then performed. The oscillator is set to 30 cycles/s. The voltage is raised from 0 to 2 V maximum (i.e. root mean square value read directly from the dial of the oscillator and not to be confused with peak voltage of an individual sine wave) in 1–2 s and remains at this maximum for about 5 s and then is reduced to zero in 1–2 s, remaining at zero for about 10 s. This sequence or "unit of stimulation" is repeated until ejaculation occurs, or for a maximum of 25 times. However, ejaculation usually occurs after the fifth or sixth stimulatory period. The ejaculate is collected either on a warmed glass plate placed beneath the rat and on which is placed a few drops of physiological solution, or directly into a suitable container.

It should be noted that the electrical values quoted above for the Wistar rat are critical, since raising the voltage maximum to 3 V or more, or the cycles per second to 60, results in a coagulated ejaculate. Moreover, the clot may stay in the urethral passage and cause death due to uraemia. Data on electrical values for other strains are not readily available, and exploratory work would have to be done by the investigator.

Electroejaculation can be performed many times using the same rat, which soon becomes conditioned and may even ejaculate spontaneously during the tying down procedure prior to the electrical stimulus. The number of sperm that can be expected in an ejaculate is about 60 million. If the sperm are to be used for artificial insemination they should be placed in a solution consisting of 7 ml distilled water, 3 g egg yolk, 15 mg sodium bicarbonate and 300 mg glucose.

5. COLLECTION OF MILK (MAMMARY GLAND SECRETION)

Quantities of milk can be obtained from lactating rats with the use of a simply constructed milking machine used under controlled conditions (Fig. 58). In addition to the components shown, a direct or geared-drive fractional horse power motor will be required to drive the metal

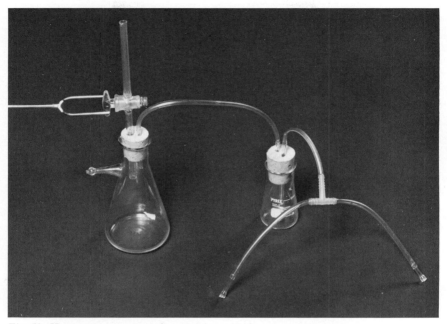

Fig. 58. Home-made apparatus for obtaining milk from lactating rats.

fork which in turn causes the stopcock to rotate. The opening in the stopcock should be aligned with the lumen of the vertical tube (10 mm o.d.) and the stopcock should then be rotated about 25 times/min. A potentiometer (rheostat) may be required to control the revolutions of the motor. A vacuum suction source will also be required, and can be supplied either by a suitable vacuum motor or by a water vacuum. The vacuum level should be between 25 and 28 cm of mercury. Because the system is "open" only 25 times/min, a series of "vacuum pulses" is set up which simulates the suckling action of the young rat. In some strains a variation from 25 pulses/min may be required to achieve a good yield of milk, and some exploratory studies may have to be done by the investigator. The teat cups are prepared from glass tubing which is "flared" at one end to fit comfortably over the nipple and to allow any extension of it during milking.

For milking, the offspring should be removed from their mother about 4 hours previously to allow the mammary glands to become engorged with milk. Thirty minutes before milking is to begin, the rat should be given a s.c. injection of 5 i.u. oxytocin to stimulate milk ejection. The teat cups are then fitted over two teats at a time and pressed against the body wall to achieve a vacuum-tight seal. A light smearing of petroleum jelly around the nipple will help to achieve this. The metal fork is then rotated and the vacuum turned on. Each gland should be massaged gently towards the nipple during the milking process. The milk may take a few minutes to start flowing and milking should be stopped when the rat becomes restless. Providing the mother has been lactating for 7 days or more, a total of 3–7 ml of milk can be expected daily at each milking. Smaller amounts of milk can be obtained by the use of an alternative and somewhat simpler apparatus (see Fairney and Weir, 1970). A few drops of milk can often be obtained merely by gently massaging the glands towards the nipples.

6. COLLECTING URINE, FAECES AND RESPIRED GASES

6.1. Simultaneous Collection of Urine and Faeces

Collection of both urine and faeces is carried out with the use of a metabolism cage. Many types of metabolism cage are available commercially (see Equipment Index), and three are shown in Fig. 59. The principle of these cages is the confinement of the rat in a cage with a wire grid floor. The cage is placed on top of a funnel device so that urine falling on the sides of the funnel is channelled out via a side arm, while the faeces pellets drop into a collecting jar. In the more sophisticated metabolism cage, feeding and watering compartments are incorporated such that food and water do not significantly contaminate the urine and faeces.

Commercially available cages can be expensive. A simple cage that can be constructed in the laboratory is shown in Fig. 60 (see also Halladay, 1973). For its construction, a five gallon (11 litre) glass carboy is etched completely round the outside with a glass cutter 29.2 cm from the base, and the glass broken along the etched line by light tapping. The top is then inverted into the bottom forming a funnel. The urine–faeces separator is made from a 50 ml Pyrex volumetric flask.

The tips of two glass rods are heated until soft and then attached to the top and bottom of the flask. The neck of the flask is then heated until soft and, by means of the attached rod, the neck is drawn apart and bent at the desired length to form a hook (100 mm from the base of the neck to the top of the hook). The rod connected to the bottom of the flask is heated and removed, forming a 0.64 mm nipple on the bottom. A 2.5 × 0.48 cm stainless steel spring is hooked onto the separator and to a 22.9 cm × 3 mm o.d. glass rod placed sideways inside the funnel, thus suspending the separator. A 50 ml glass beaker is used to collect the urine, and a sheet of aluminium foil collects the faecal pellets which bounce off the separator.

The cage to fit over the funnel is fashioned from 1.3 × 1.3 cm wire mesh sheet. The dimensions of the top and floor are 30.5 × 30.5 cm, with a 1.3 cm 90° bend on each side and a 14 × 27.9 cm hinged door incorporated into the top. The sides are 22.9 × 82.6 cm arranged in a cylinder with a diameter of 26.7 cm and secured to the floor and top with pieces of wire.

If a glass carboy is unobtainable, a suitable glass or plastic container of similar size, and a large glass, metal or plastic funnel suitably modified to accommodate the urine–faeces separator could be used as a substitute, given a little ingenuity on the part of the investigator.

Experiments requiring the measurement of respired gases (e.g. CO_2) also make use of metabolism cages, and one such commercially available apparatus is shown in Fig. 61. The basic central unit of this apparatus can also be used solely for the collection of urine and faeces if required.

6.2. Collection of Faeces

Although metabolism cages allow the separation of urine and faeces, there is nevertheless some contamination of one by the other, and perhaps by food and water. Collection of uncontaminated faecal matter can be done by employing an anal cup. Because the distance between the anal and urethral orifices of a female rat is nearly half that of a male rat, it is preferable to use anal cups only with male rats.

One design for an anal cup is shown in Fig. 62 (see also Ryer and Walker, 1971). For its construction, the bottom of a long narrow-mouthed flexible polyethylene bottle (approximately 2.5 cm diameter with a 1.3 cm diameter mouth) is cut off at a 60° angle with a scalpel.

(A)

Fig. 59. Commercially available metabolism cages.

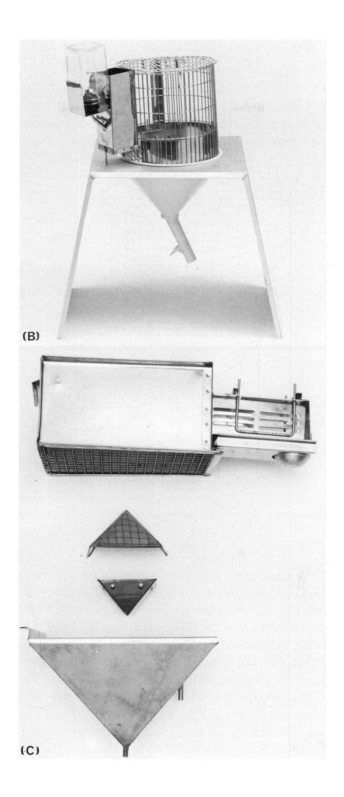

(B)

(C)

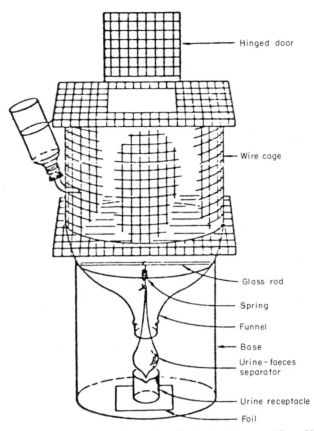

Fig. 60. Metabolism cage which can be made easily in the laboratory (from Halladay, 1973).

For a 200–250 g rat, the upper portion should be 7.5 cm long and the lower part 6.4 cm long. For a 100–150 g rat the relative proportions are decreased by 1.3 cm, and even more for smaller rats. Using scissors a U-shaped curve is cut into the lower shorter side of the cup, the depth of the U being determined by the anogenital distance of the rat to be fitted, and therefore determined by a process of trial and error. The cup should now fit snugly against the posterior of the rat without causing irritation of the sensitive genital area nor allowing urine to enter.

For positioning, the rat is held securely and the anal cup is slipped over the tail and against its posterior. The longer portion should be uppermost and dorsal while the shorter side with the U should be

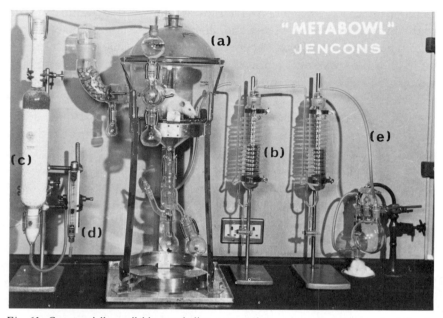

Fig. 61. Commercially available metabolism cage and accessory apparatus required for the measurement of respired gases ("Metabowl", Jencons Ltd): (a) glass metabolism cage, (b) two nilox columns in series containing 4 N sodium hydroxide solution, (c) air drier and carbon dioxide absorber column containing Drierite below and soda lime above, (d) flowmeter to allow air at a rate of 0.25–0.4 1/min, (e) tubing to Cartesian manostat and then to vacuum line.

Fig. 62. Materials required for the preparation of an anal cup (from Ryer and Walker, 1971).

carefully positioned to just cover the anal opening. A 10.2 cm length of 2.5 cm wide adhesive tape is wound round the neck of the bottle leaving about 1.3 cm extending beyond the neck to provide a flexible connection to the tail. A 15.2 cm length of 1.3 cm wide tape is used to connect the protruding portion of the wider tape to the tail of the rat. The narrow tape is wound round both the protruding tape and the rat's tail several times to ensure the cup is held firmly against the body. The rat can now be returned to its cage, preferably one with a grid floor, or a metabolism cage. The collected faeces are removed by squeezing the anal cup and drawing the open end slightly away from the body on one side. The faecal pellets are "spooned" out and the anal cup is returned to its normal position. Since defaecation is most copious at night, removal of faeces is best carried out first thing in the morning.

A distinct advantage of using an anal cup, especially in nutrition studies, is that coprophagy is prevented. Coprophagy not only delays the ultimate excretion of dietary compounds but may alter the quantitative utilisation of a nutrient. Moreover, the gut flora may be changed as a result of coprophagy which may then affect the way in which dietary materials are digested. It is estimated that rats, particularly in cages with wire grid floors, may ingest 50–65% of the faeces produced.

6.3. Collection of Urine

If rats wearing anal cups are placed in a metabolism cage, urine uncontaminated with faeces can be collected. However, for collection of small amounts of urine, it is often sufficient to catch and remove an animal from its cage and rapidly position its urethral opening over a collecting jar or tube. Rats very often urinate on being handled. However, if rats are handled and familiarised frequently then spontaneous urination may not occur. In such circumstances an ether-or methoxyflurane-soaked pad placed over the rat's nose will usually cause it to struggle and to urinate. This manoeuvre also sometimes works in animals which have urinated before the urine could be collected. Stressing the animal in this way may produce a little "extra" urine, especially if the abdominal area over the bladder is pressed with the fingers.

7. COLLECTION OF OTHER BODY FLUIDS

7.1. Collecting Bile and Pancreatic Secretions

See Sections 20.1 and 20.2.

7.2. Collecting Gastric Contents

See Section 23.

7.3. Collecting Cerebrospinal Fluid

See Section 2.18.

7.4. Collecting Lymph

See Section 28.

Chapter Three

ANAESTHESIA AND POSTOPERATIVE CARE

8. ANAESTHESIA—AIMS AND OBJECTIVES

Anaesthesia of laboratory animals aims to provide humane restraint, a reasonable degree of muscle relaxation so that surgery can be carried out easily and most importantly, a sufficient degree of analgesia to prevent the animal experiencing pain.

In selecting a method of anaesthesia a number of other points should be considered. If anaesthesia is intended to provide humane restraint, then the method of administration should cause a minimum of distress to the animal. Techniques such as induction with an irritant volatile anaesthetic such as ether can be highly distressing for the animal, particularly if the rat is placed in a badly designed induction chamber and comes into contact with the liquid anaesthetic. Injection of either large volumes of drugs intramuscularly, or injection of an irritant compound, can cause unnecessary distress to the rat. Whatever method of anaesthesia is chosen, careful and expert handling of the animal, with consideration of ways of minimising the fear and stress associated with movement from the animal holding room to the operating theatre, is also important.

Aside from these preoperative considerations, high standards of intraoperative and postoperative management must be maintained if the animal is to make an uneventful recovery from anaesthesia. Poor intraoperative care can result in very prolonged recovery times. This can lead to problems such as injury of semiconscious animals by cage mates, or damage to the eyes from abrasion by sawdust or other bedding. Rats may also develop skin irritation following soiling of the fur with urine and faeces. It seems reasonable to assume that problems such as these, and also the prolonged inappetence that can result following poor anaesthetic management, are distressing to the animal and so are best avoided. The prevention of postoperative pain is particularly important, and is discussed in detail later in this chapter.

Aside from the importance to animal welfare of good anaesthetic practice, clearly most of these considerations are also relevant to the scientific validity of any study which makes use of the animals. Most research workers strive to produce an animal model which is carefully defined, and has the smallest degree of unwanted variability. If the rat is to be allowed to recover following experimental surgery, then it should return to physiological normality, or to a defined state of abnormality, as rapidly as possible. These scientific aims are easily frustrated by poor anaesthetic practice. Pain, fear and distress can become uncontrolled variables which interfere with a research protocol. Similarly, a rat that does not eat or drink normally for 24–48 hours, or which develops severe hypothermia, or respiratory acidosis or hypoxia, can hardly be considered a good animal model.

There remains one further important objective when anaesthetising laboratory rats, which is to select an anaesthetic technique which interferes as little as possible with the particular experiment which is being undertaken. Achieving this aim is difficult, but a careful assessment of the alternative anaesthetics that are available, and their particular physiological and pharmacological effects, can at least minimise the interaction between anaesthetic regimen and the particular animal model. It must be appreciated that this type of assessment will not always have been carried out by research workers whose published accounts of their research includes details of the anaesthetic regimen used. Simply adopting the method of anaesthesia described in publications dealing with the particular research procedure will not therefore necessarily ensure that an appropriate technique is used. A brief review of the most important features of individual anaesthetics is included later in this section.

Response to surgical stress: when evaluating the potential interactions between the anaesthetic regimen and the experimental protocol, it is important to consider the effects of anaesthesia together with the influences of the surgical procedure. Surgery produces a stress response whose magnitude is related to the severity of the operative procedure. The surgical stress response in mammals consists of a mobilisation of reserves of substrates, for example glucose, to enable the organism to survive injury. This response has clear evolutionary advantages, but it is believed by many to be undesirable and inappropriate in humans and animals which are receiving a high level of intraoperative and postoperative care (Hall, 1985; Salo, 1988).

These major responses to surgery are characterised in the rat by an elevation in plasma concentrations of catecholamines, corticosterone, growth hormone, vasopressin, renin, aldosterone and prolactin, and by a reduction in plasma concentrations of follicle stimulating hormone, luteinising hormone and testosterone. The effects of surgery on plasma insulin and glucagon are more complex. Initially insulin concentrations decrease and glucagon increases, but later insulin concentrations rise. These hormonal responses to tissue trauma produce an increase in glycogenolysis and lipolysis, and result in hyperglycaemia. The duration of the hyperglycaemia varies, but after major surgery the response may persist for 4–6 hours. More prolonged changes in protein metabolism occur leading to negative nitrogen balance lasting for several days (Hoover-Plow and Clifford, 1978). Even relatively minor surgical procedures can produce relatively prolonged effects. For example, blood vessel catheterisation in rats has been shown to produce an elevation of corticosterone for several days (Fagin et al., 1983), and more subtle changes such as disruption of the circadian rhythmicity of hormonal secretions can also persist for similar periods (Desjardins, 1986).

Research workers are often reluctant to refine their anaesthetic methodology because it is perceived that the anaesthetics used in the improved regimen may have an effect on their animal model in the postsurgical period. In some instances there are sound scientific reasons for avoiding certain agents, but what is often not appreciated is that in many instances the effects of anaesthesia are overshadowed by the major stress effects of surgery. Similar concerns are often expressed that administration of analgesics could cause an alteration to the animal model which is to be studied. Once again, the side effects of analgesics are often masked by the changes induced by surgical trauma, and in some instances controlling postsurgical pain may reduce the magnitude of the metabolic response to surgery. Clearly it is logical to consider all of the factors that may interact with a particular study, and develop an anaesthetic and surgical protocol which is both humane and provides the minimum of interference with the overall aims of the research project.

There remains one further problem when changing anaesthetic regimens or trying to use a method described in a publication. The response of rats to anaesthetics varies very considerably between different strains (Lovell, 1986). A dose rate which anaesthetises one strain of rat may be ineffective in another. For example, in the authors' experience, fentanyl/fluanisone with midazolam provides excellent surgical anaesthesia in Wistar rats, but in DA rats the combination fails to do so and increasing the dose rate can cause respiratory failure. Similar problems seem to occur with many other injectable anaesthetics in this strain, and it seems reasonable to suggest that when anaesthetising DA rats, a volatile anaesthetic should be used whenever possible. The variations in the response to different anaesthetics in different strains of rat have not been well characterised. An additional complication is the variation which occurs between males and females, and that caused by differences in the animal's environment. For example rats kept on soft wood beddings have increased liver microsomal activity and so have a reduced sleep time after administration of pentobarbitone. There is also a small variation in response depending upon the time of day when the anaesthetic is administered.

Because of the potentially large variations that can be observed it is advisable to begin by anaesthetising only a single animal, assess the depth of anaesthesia very carefully, record the duration of anaesthesia, and monitor the animal for several days to ensure that recovery has been complete, with no adverse postanaesthetic effects. In some instances, for example when using chloral hydrate for the first time, monitoring should be continued for 7–10 days because of the possibility of gut disturbances occurring following anaesthesia. After establishing that the anaesthetic regimen is safe and effective, it can be used with larger numbers of animals. By adopting this cautious approach, disasters caused by variation in response between rats of different strains will be much less likely to occur.

8.1. Selecting an Anaesthetic Technique

When faced with the long list of anaesthetic drugs and drug combinations which have been recommended for use in rats, it is very tempting to take the simple approach of using only one or two methods. Although this has the advantage of standardising anaesthetic methodology, it is most unlikely that the technique used will be the one that is best suited to each type of experiment. Unless only one procedure is being undertaken, regimens which provide anaesthesia of different durations and different depths will be required. In addition, some anaesthetic drugs will be contraindicated in certain experimental protocols, whilst others will be unsuitable in different circumstances.

Leaving aside the problem of anaesthetic/experiment interaction, this section attempts to review the most important points you should consider when selecting a particular anaesthetic regimen, and describes the main features of the different anaesthetics which are available.

8.1.1. Route of Administration

The choice between administering an anaesthetic by inhalation or by injection is often the first decision that is made by an investigator. Immediately, one large group of potentially useful techniques has been discarded. Rather than regard these techniques as mutually exclusive, consider not only using either an injectable anaesthetic or an inhalational technique, but also a combination of both methods. Apart from limiting the selection of anaesthetic agents, if an injectable anaesthetic regimen is chosen it is often presumed that an anaesthetic machine need not or cannot be used. The anaesthetised rats will therefore not receive oxygen supplementation, and so may become severely hypoxic.

8.2. Inhalational Anaesthetics

8.2.1. Anaesthetic Apparatus

If the correct apparatus is available, inhalational anaesthetics have the advantage of being easy to administer. Although it is possible to put volatile anaesthetic on a cotton wool swab, place it in a glass jar and drop in the animal, such an anachronistic anaesthetic technique has virtually nothing to recommend it except that it is inexpensive. Every research laboratory should invest in an anaesthetic machine and purchase or construct an anaesthetic induction chamber (Cooper, 1982) (Fig. 63, see Equipment Index). The chamber should have transparent sides, so that the animal can be observed during induction, and should also be easy to clean, otherwise it will soon become heavily soiled with urine and faeces. If a commercially produced anaesthetic trolley cannot be purchased because of budgetary constraints, consider constructing a simple apparatus from commercially available components. A laboratory trolley can be used to mount an oxygen cylinder fitted with a pin-index regulator and flowmeter. This can be connected to a suitable vaporiser, and then to an induction box or face mask (Fig. 64). Because of the risk to human health from exposure to waste anaesthetic gases, some form of gas scavenging should be fitted (Fig. 65).

Aside from the anaesthetic machine and induction chamber, anaesthetic circuits and connecting tubing will also be required. A wide range of "disposable" apparatus is now produced for human use, and this can often be modified to provide inexpensive circuits and connecting tubing.

Fig. 63. Anaesthetic induction chamber.

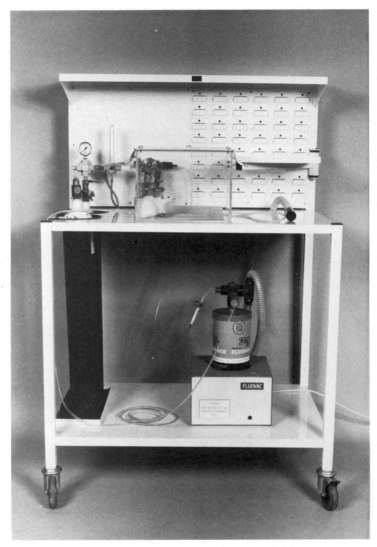

Fig. 64. Anaesthetic workstation. Oxygen cylinders are fitted with combined pin-index regulator and flowmeter. Gas scavenging is provided using the "Fluo-Vac" system (see Equipment Index).

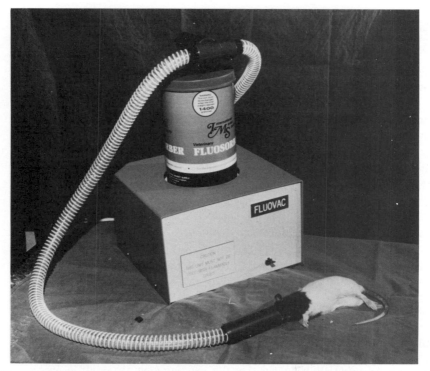

Fig. 65. Maintenance of anaesthesia using a face mask. Anaesthetic gases are delivered through a central tube and mask, and gas scavenging carried out by means of a concentrically mounted outer mask and vacuum suction.

8.2.2. *Induction and Maintenance of Anaesthesia with Inhalational Agents*

Induction of anaesthesia is straightforward, but many investigators are a little daunted by the apparatus at first. It is useful to provide a step-by-step guide for new investigators, which can be attached or displayed beside the apparatus. The guide shown in Table 5 is used in the author's Institute, and can easily be modified for use with different apparatus.

The use of an anaesthetic induction chamber is probably the easiest and least stressful means of inducing anaesthesia with a volatile anaesthetic (Fig. 63). A layer of paper towels should be placed on the floor of the chamber to absorb any urine and ease cleaning. All of the commonly used volatile anaesthetics (halothane, methoxyflurane,

Table 5. Instruction for use of a rodent anaesthetic apparatus

1. Turn on the oxygen cylinder using the key provided. The pressure gauge on the regulator should register a pressure above zero. Check the level of liquid anaesthetic in the vaporiser and ask for the vaporiser to be refilled if necessary.

2. Turn on the flow meter (mounted on top of the cylinder), allow the oxygen to run for about 30 seconds, and ensure that the cylinder pressure gauge has not dropped to zero.

3. Check that the pipe line from the halothane vaporiser is connected to the induction chamber, and that the gas scavenging device is switched to scavenge from the chamber.

4. Place a rat onto a paper towel in the bottom of the chamber.

5. Turn the vaporiser setting to 3%, and the oxygen flowmeter to 1 l/min—the reading is taken from the top of the bobbin.

6. When the rat has lost its righting reflex and is sleeping quietly, switch the gas scavenging system and vaporiser connections to the face mask. Remove the rat, place its nose in the face mask, and replace the lid of the induction box. Reduce the vaporiser setting to 1.5 or 2%. Rats vary slightly in their response to halothane, but 1.5% is usually sufficient.

7. Repeat steps 3 and 4 as necessary. Once you are familiar with halothane, you will find that placing a second rat in the induction chamber as you remove the first one will speed up the process. You can also increase the speed of anaesthetic induction by using 4% halothane, but this concentration can result in overdose and death.

8. After completion of the surgery, place the rat in a cage, on "Vetbed"—NOT sawdust—in the incubator at 37°C. It should recover in 5–10 minutes and can then be returned to its cage in the animal holding room.

9. At the end of the procedure, turn off the vaporiser, flowmeter and the oxygen cylinder. Turn off the gas scavenging system. Remove the paper towel from the induction box and clean the induction chamber. Clean the surface of the rodent trolley.

Problems

If a rat starts to gasp and make violent respiratory movements, or if its respirations become very shallow, it is too deeply anaesthetised. Remove it from the mask or chamber until it begins to breath normally, then continue but with a reduced concentration of halothane.

If a rat stops breathing, remove it from the mask or chamber, lay it on its back with its head extended and gently squeeze its chest several times between your thumb and forefinger. When it starts breathing, wait a minute or so for the pattern to stabilise, then put it back on the mask.

enflurane and isoflurane) produce smooth induction of anaesthesia. Induction is slower with methoxyflurane, but this can be an advantage for inexperienced anaesthetists. A flow of about 1–2 l/min of oxygen should be used with a small induction chamber (e.g. 30 cm × 20 cm × 20 cm), together with the recommended induction concentration (Table 6) of the anaesthetic agent, delivered until the rat loses its righting reflex and remains immobile in the chamber. This will take around 2–3 min when using halothane. The animal can then be removed from the chamber and a face mask used to maintain anaesthesia (Fig. 65). It is important to remember to reduce the concentration of anaesthetic to the recommended maintenance concentration at this stage. If a brief procedure is to be carried out which will not require anaesthesia to be maintained using a face mask, the rat can simply be removed from the chamber and the procedure carried out. Anaesthesia will persist for 30–90 seconds, full recovery of consciousness will take a few minutes. If an anaesthetic machine is unavailable, it is possible to use methoxyflurane in a simple "anaesthetic jar" (see below and Fig. 66). When gaining experience with the use of volatile anaesthetics, once the animal has lost its righting reflex, it may be found helpful to reduce the concentration of anaesthetic delivered to the induction chamber to the recommended maintenance concentration for a few minutes. This will allow the animal to reach a stable level of anaesthesia and will provide a slightly longer period of anaesthesia after removal from the induction chamber. Once experience has been gained with a particular anaesthetic agent then this short period of maintenance can be eliminated, and the rat simply removed from the chamber when the required depth of anaesthesia has been obtained. Deepening of anaesthesia is usually indicated by a change in respiratory pattern and a reduction in respiratory rate. If animals are required to be euthanased at the end of a study,

Table 6. Concentrations of volatile anaesthetic agents for use in rats.

Anaesthetic	Induction concentration %	Maintenance concentration %
Enflurane	3–5	0.5–2
Ether	10–20	4–5
Halothane	3–4	1–2
Isoflurane	3.5–4.5	1.5–3
Methoxyflurane	4	0.4–1

Fig. 66. Simple anaesthetic chamber for use with methoxyflurane. The liquid anaesthetic (5 ml) is placed on cotton wool held in a stainless steel coffee strainer attached to the lid. The cotton wool in the jar is provided for the animal's comfort. Note that effective gas scavenging is extremely difficult to achieve when using this type of apparatus.

then when this is carried out it can be used as an opportunity to gain experience with the signs of both surgical anaesthesia and anaesthetic overdose.

The most important requirement for successful anaesthesia using volatile anaesthetics is careful observation of the rat during induction. In our experience, the most common cause of inadvertent overdose is when the role of anaesthetist and surgeon are combined, and the operator attempts to prepare instruments or other equipment whilst simultaneously inducing anaesthesia. If combining the duties of anaesthetist and surgeon are unavoidable, then especially careful planning and attention to detail are required if unnecessary anaesthetic mortality is to be avoided.

If anaesthesia has been maintained for 20–30 minutes, then recovery is usually complete in 5–10 minutes, but after several hours of anaesthesia full recovery can take 15 or 20 minutes or even longer. The speed of recovery will vary considerably depending upon the attention which has been given to controlling anaesthetic depth by varying the concentration of anaesthetic agent delivered to the animal. Generally, as anaesthesia progresses, the concentration of anaesthetic can be reduced slightly. Following the completion of major surgery, a further reduction in depth can be made during suturing of the subcutaneous tissues and skin, so that recovery time following completion of surgery is reduced. This ability to rapidly vary the depth of anaesthesia is one of the major advantages of using inhalational agents.

8.2.3. Inhalational Agents Available

8.2.3.1. *Halothane.* Halothane is a potent anaesthetic which is easy to vaporise and is non-flammable and non-explosive. Because of its potency, it should only be used in a calibrated vaporiser, so that the concentration delivered to the animal can be carefully controlled. Halothane depresses the cardiovascular and respiratory systems in a dose-related manner. In healthy rats these effects are not usually clinically significant at normal maintenance concentrations. During recovery from anaesthesia, halothane is exhaled from the lungs, but significant quantities of this anaesthetic are metabolised in the liver and a postanaesthetic increase in liver microsomal enzymes occurs. Liver enzyme induction is likely to be significant only after prolonged periods of anaesthesia (>30–60 min), but detectable effects on drug metabolism both during and after anaesthesia have been reported (Linde and Berman, 1971; Wood and Wood, 1984). Halothane is pleasant to inhale and non-irritant to the respiratory system, and it is an excellent anaesthetic for use in the rat.

8.2.3.2. Enflurane. Enflurane resembles halothane in many of its characteristics, but induction and recovery are slightly more rapid. It undergoes less hepatic metabolism than halothane, and can be used to provide safe and effective anaesthesia in the rat.

8.2.3.3. Isoflurane. Isoflurane produces even more rapid induction of and recovery from anaesthesia than does halothane, but like halothane it also causes some cardiovascular and respiratory system depression. It has a more pungent odour than halothane, but is non-irritant and its administration does not appear to be resented by rats. Isoflurane is non-flammable and non-explosive, and undergoes virtually no hepatic metabolism (Eger, 1981). Because of its potency, it must be administered using a calibrated vaporiser. In many respects it could be considered the most suitable volatile anaesthetic for use in the rat, but its high cost relative to halothane, coupled with the additional cost of a calibrated vaporiser, have restricted its use in research animal units.

8.2.3.4. Methoxyflurane. Methoxyflurane can be delivered either using a calibrated vaporiser, or using a simple "bubble through" system such as a Boyles bottle. It can also be used directly on a cotton wool swab in an anaesthetic chamber, provided that there is no possibility of the animal coming into contact with the liquid anaesthetic (Fig. 66). Induction of anaesthesia is slow relative to halothane, and this can be an advantage for the inexperienced anaesthetist since it allows a greater margin of safety when judging anaesthetic depth. This broader safety margin makes it an excellent anaesthetic for neonatal animals, in which assessment of depth of anaesthesia is often difficult. Methoxyflurane produces some respiratory and cardiovascular system depression, but these effects are generally slightly less than that caused by halothane at comparable depths of anaesthesia. Metabolism of methoxyflurane results in some inorganic fluoride ion release, and after prolonged anaesthesia this can result in kidney damage (Murray and Fleming, 1972).

8.2.3.5. Nitrous Oxide. Nitrous oxide cannot be used to produce either surgical anaesthesia or loss of consciousness in the rat without causing severe hypoxia. The anaesthetic potency of volatile anaesthetics is generally expressed as the minimum alveolar concentration (MAC) of an agent that will prevent a purposive movement to a defined noxious stimulus, usually a haemostat clamped on the tail or a digit. The MAC_{50} concentration for nitrous oxide in the rat is 150%, so it is impossible to provide surgical anaesthesia using nitrous oxide alone. Attempts to induce anaesthesia with nitrous oxide in the rat fail to produce loss of consciousness until severe anoxia has been produced by delivering more than 95% nitrous oxide. For this reason, attempts to maintain anaesthesia with nitrous oxide, together with a muscle relaxant to paralyse the animal and prevent it moving, must be considered inhumane and unacceptable. Nitrous oxide does, however, have some analgesic action, and it can be used in combination with other agents which produce loss of consciousness as part of a balanced anaesthetic regimen. During prolonged periods of anaesthesia (>24 h) it may be undesirable to allow the animal to inhale 100% oxygen because of the risk of development of oxygen toxicity. Administration of a mixture of 60% nitrous oxide and 40% oxygen will overcome this problem. Use of this gas mixture will also allow the concentration of other volatile agents such as halothane to be slightly reduced. Since nitrous oxide has little effect on the cardiovascular or respiratory systems, this can be advantageous.

8.2.3.6. Ether. Ether is flammable and when volatilised it produces an irritant vapour which forms explosive mixtures with air or oxygen. Because of the proven dangers of ether explosions and fires, many research units have abandoned the use of this anaesthetic because of the safety hazard that it represents. Aside from this safety hazard, ether is unpleasant to inhale, and its irritant nature causes an increase in airway secretions and can exacerbate any pre-existing subclinical respiratory disease. It has remained a popular anaesthetic in some institutes because it can be used directly in simple anaesthetic chambers. Like methoxyflurane, its slow induction rate provides a considerable margin of safety for the inexperienced anaesthetist. Although ether has a depressant effect on the myocardium, it also stimulates the sympathetic nervous system and the overall effect on the circulation is to maintain blood pressure. Ether undergoes hepatic metabolism and causes microsomal enzyme induction (Linde and Berman, 1971). Because of the safety hazard and the unpleasant nature of induction with ether, it should be used only under very exceptional circumstances.

8.3. Injectable Anaesthetics

8.3.1. Route of Administration

Most injectable anaesthetics are administered to rats by the i.p. or i.m. routes as a single dose. Each route is claimed to have some advantages, for example intraperitoneal administration enables larger volumes of anaesthetic to be administered, and the technique, when carried out skilfully, appears to cause a minimum of pain and distress to the rat. It is possible, however, inadvertently to administer anaesthetic into the lumen of the intestine, or into the bladder or the subcutaneous fat. Such failures of intraperitoneal drug delivery will be rare once experience has been gained, but intramuscular administration of small volumes of anaesthetic into the quadriceps muscle may provide more reliable absorption and drug delivery. Unfortunately, some anaesthetics have been shown to produce muscle damage when administered by this route (Smiler *et al.*, 1990), and in general the authors prefer the intraperitoneal route. Both methods of administration share a major disadvantage, in that a large single dose of anaesthetic must be given to produce the required effect. Absorption is slow, residual drug effects persist for long periods, and so full recovery can be very prolonged.

In larger species, administration by the intravenous route enables dosing to be adjusted according to the individual animal's response, and so inadvertent over- or underdosing is easy to avoid. Intravenous administration has obvious advantages in terms of control of depth of anaesthesia, and in addition enables a wider range of anaesthetic agents to be used. It is only the perceived difficulties, such as restraint and the technical problems of i.v. administration, that prevent widespread adoption of this technique in smaller species such as the rat. Research workers will often cite these problems as reasons for avoiding intravenous administration of anaesthetics, but they may routinely administer other substances by the i.v. route as part of a research protocol! Before discounting i.v. administration, consider whether the necessary technical expertise is present amongst personnel in the research laboratory, or whether the effort of developing this expertise would be worthwhile. A number of anaesthetics such as propofol, alphaxalone/alphadolone, methohexitone and thiopentone (see 8.3.2.1.) can all be used to produce short periods (5–10 minutes) of anaesthesia, followed by rapid recovery, and some can be used to produce prolonged anaesthesia when administered by continuous i.v. infusion. Injection techniques and recommendations for needle sizes and acceptable volumes are described in Chapter One.

8.3.2. Injectable Agents Available

A wide range of injectable anaesthetics are available for use in rats. The duration and quality of anaesthesia produced by different agents varies considerably, and is summarised below. Although the degree of analgesia produced by some compounds is inadequate for major surgery, addition of a low concentration of volatile anaesthetic to the regimen can provide improved levels of analgesia.

8.3.2.1. Short duration anaesthesia (up to 10 minutes). Propofol: propofol (10 mg/kg) produces approximately 5 minutes surgical anaesthesia when administered intravenously in the rat. Recovery is smooth and rapid, and administration of repeated injections does not unduly prolong recovery (Glen, 1980). Propofol is a modified phenol, and is prepared in a fat emulsion for intravenous injection. It causes a mild degree of hypotension and moderate respiratory depression. High doses can cause respiratory arrest, but since the drug is rapidly metabolised, spontaneous respiration resumes within a few minutes provided that ventilation is manually assisted. When administered by continuous intravenous infusion, more prolonged periods of anaesthesia are produced. Cardiovascular and respiratory depression are generally mild and recovery following cessation of the continuous infusion is rapid (Glen, 1980). Because of the rapid redistribution and metabolism of propofol, it cannot be administered effectively by the intramuscular or intraperitoneal routes.

Alphaxalone/alphadolone: this mixture of steroids is prepared in a solubilising agent (Cremophor EL) which promotes histamine release in some species. In the rat this does not appear to be a significant problem, and alphaxalone/alphadolone (10–15 mg/kg) produces approximately 5 minutes of surgical anaesthesia with a good degree of muscle relaxation when administered intravenously (Green et al., 1978). Respiratory and cardiovascular system depression is generally minimal, and prolonged periods of stable anaesthesia can be produced if this anaesthetic is administered by continuous infusion. Recovery following cessation of the infusion is rapid, and animals regain their righting reflex in 20–30 minutes. Administration of repeated doses of this anaesthetic does not significantly extend the recovery time, which is smooth and rapid. Because of the rapid redistribution and metabolism of alphaxalone/alphadolone, it cannot be administered effectively by the intraperitoneal routes. The volume of anaesthetic required to induce anaesthesia by intramuscular injection is too great to be of any value.

Thiopentone: this short acting barbiturate produces approximately 10 minutes surgical anaesthesia when administered intravenously (30 mg/kg). Moderate cardiovascular and respiratory depression are produced, but recovery is rapid, although not as smooth as with propofol. Administration of additional doses of this anaesthetic prolongs the period of anaesthesia, but also greatly increases the recovery time. This agent is ineffective when administered by the intraperitoneal or intramuscular routes.

Methohexitone: like thiopentone, a short period of anaesthesia (5 min) is produced following intravenous injection (7–10 mg/kg). Moderate cardiovascular and respiratory depression are produced. Recovery is rapid, but associated with excitement. Anaesthesia can be prolonged by administration of up to two additional doses without unduly prolonging the recovery period. This agent is ineffective when administered by the intraperitoneal or intramuscular routes.

8.3.2.2. *Medium duration anaesthesia (up to 1 hour)*. *Fentanyl/ medetomidine*: fentanyl is a potent μ opioid which provides excellent analgesia in the rat. Medetomidine is a recently produced alpha$_2$-adrenergic agonist which produces loss of consciousness and some degree of analgesia. When combined (250 μg/kg fentanyl and 300 μg/kg medetomidine, mixed and administered as a single i.p. injection), these two compounds produce surgical anaesthesia which lasts for approximately 45 minutes. Both agents have significant side effects which are discussed below. Recovery, assessed as the return of the rat's righting reflex, is very prolonged (>120 minutes), but the effects of this drug combination can be completely reversed at any stage by the administration of the specific antagonists nalbuphine (2 mg/kg i.p. or s.c.) together with atipamezole (1 mg/kg i.p. or s.c.) (Hu *et al.*, 1992).

Ketamine: when used as the sole anaesthetic (100 mg/kg, i.p.), rats are immobilised but insufficient analgesia is provided for even the most superficial surgical procedure. Ketamine stimulates the sympathetic nervous system and causes an elevation of heart rate and blood pressure. When combined with other compounds, ketamine produces light to medium planes of surgical anaesthesia.

Ketamine/xylazine: this is the most widely used ketamine combination (90 mg/kg + 10 mg/kg i.p.). The addition of xylazine, an alpha$_2$-adrenergic agonist with sedative and analgesic properties, results in medium planes of surgical anaesthesia (Wixson et al., 1987). Despite some claims to the contrary, the hypotensive effects of xylazine predominate and this combination produces a fall in blood pressure. Respiration is depressed and recovery is prolonged (Table 7). Xylazine produces hyperglycaemia, and has a very marked diuretic effect. Recovery time can be shortened dramatically by the administration of atipamezole (1 mg/kg i.p.), an alpha$_2$-antagonist which, unlike yohimbine, appears devoid of undesirable side effects.

Ketamine/medetomidine: medetomidine has been reported to have less marked side effects than xylazine, and to have a more pronounced analgesic effect. When used in combination with ketamine (75 mg/kg + 0.5 mg/kg) it produces moderate surgical anaesthesia (Nevalainen et al., 1989) although it has yet to be used in a wide range of different strains of rat. Medetomidine, like xylazine, produces a marked hyperglycaemia and diuresis. Recovery is prolonged (Table 7) but can be reduced considerably by the administration of atipamazole (1 mg/kg).

Ketamine/acepromazine: the combination of ketamine and acepromazine (75 mg/kg + 2.5 mg/kg) (Farris and Snow, 1987), a phenothiazine tranquillizer, produces light surgical anaesthesia in most strains of rat. In some strains a depth of anaesthesia sufficient for major surgery such as laparotomy is achieved. Blood pressure is reduced, primarily because of the peripheral vasodilation of the acepromazine and moderate respiratory depression occurs.

Ketamine/azaperone: ketamine (45 mg/kg) and azaperone (25 mg/kg) have been reported to produce surgical anaesthesia in the rat (Olson and Renchko, 1988). In the authors' experience the degree of analgesia is very variable.

Tiletamine: when used alone this dissociative anaesthetic, like ketamine, does not produce even light anaesthesia. It is marketed commercially in combination with a benzodiazepine, zolezapam (Telazol, Parke Davis). This combination produces light to medium plane anaesthesia (40 mg/kg i.p.) (Silverman et al., 1983) but there appear to have been few reports of its effects on the major body systems.

Neuroleptanalgesics: Several commercial preparations are available which combine a potent opioid analgesic with a phenothiazine or butyrophenone tranquilliser.

Table 7. Anaesthetic and other compounds for use in the rat. Considerable variation in response between different strains of rat can be anticipated. Always undertake a pilot study when changing to a new anaesthetic regimen. All dose rates are for i.p. injection unless otherwise stated.

Compound	Dose rate	Comments	Duration of anaesthesia	Sleep time[*]
Acepromazine	2.5 mg/kg	Moderate sedation	—	—
Alphaxalone/ alphadolone	10–15 mg/kg	Surgical anaesthesia only if i.v.	5 min	10 min
Atropine	0.05 mg/kg	Give to reduce salivation	—	—
Chloral hydrate	400 mg/kg	Postanaesthetic ileus may occur	60–120 min	2–3 h
Chloralose	55–65 mg/kg	Light/moderate prolonged anaesthesia, non-recovery only	8–10 h	—
Diazepam	2.5 mg/kg	Moderate sedation	—	—
Fentanyl/ medetomidine	250 μg/kg 300 μg/kg	Surgical anaesthesia	45 min	3–4 min
"Hypnorm" (fentanyl/fluanisone)	0.4 ml/kg i.m.	Surgical analgesia, immobilisation, poor muscle relaxation	20 min	60 min
"Hypnorm" (fentanyl/fluanisone) + diazepam	0.3 ml/kg i.m. 2.5 mg/kg	Surgical anaesthesia	45–60 min	2–4 h
"Hypnorm" (fentanyl[†]/fluanisone) + midazolam	2.7 ml/kg[†]	Surgical anaesthesia	45–60 min	2–4 h
Immobilon SA (etorphine/ methotrimeprazine)	0.5 ml/kg i.m.	Surgical analgesia	60 min	2–4 h
Inactin	80 mg/kg	Variable duration and depth of anaesthesia	1–4 h	2–3 h

Table 7. Continued.

Compound	Dose rate	Comments	Duration of anaesthesia	Sleep time*
Innovar Vet (fentanyl/droperidol)	0.4 ml/kg i.m.	Analgesia/ immobilisation, poor muscle relaxation	20–30 min	1–2 h
Ketamine	100 mg/kg i.m.	Sedation/ immobilisation	20–30 min	2 h
Ketamine/ acepromazine	75 mg/kg 2.5 mg/kg	Light anaesthesia	20–30 min	2 h
Ketamine/ diazepam	75 mg/kg 5 mg/kg	Light anaesthesia	20–30 min	2 h
Ketamine/ medetomidine	75 mg/kg 0.5 mg/kg	Surgical anaesthesia	20–30 min	2–4 h
Ketamine/ xylazine	90 mg/kg 10 mg/kg	Surgical anaesthesia	20–30 min	2–4 h
Methohexitone	10 mg/kg i.v.	Surgical anaesthesia only if i.v.	5 min	10 min
Pentobarbitone	40 mg/kg 60 mg/kg	Narrow safety margin	15–60 min	2–4 h
Propofol	10 mg/kg i.v.	Surgical anaesthesia only if i.v.	5 min	10 min
Thiopentone	30 mg/kg i.v.	Surgical anaesthesia only if i.v.	10 min	15 min
Tiletamine/ Zolezepam	40 mg/kg (of commercial preparation)	Light/moderate surgical anaesthesia	15–25 min	1–2 h
Urethane	1000 mg/kg	Prolonged surgical anaesthesia, non-recovery only	6–8 h	—
Xylazine	10 mg/kg	Sedative/some analgesia	—	—

*Sleep time is duration of loss of righting reflex.
†Mixture of one part "Hypnorm", two parts water for injection and one part midazolam.

Fentanyl/fluanisone ("Hypnorm"): when used as the sole anaesthetic agent (0.4 ml/kg i.m.), potent analgesia is produced, sufficient to enable surgery to be undertaken, but this is associated with poor muscle relaxation and pronounced respiratory depression. These disadvantages can be overcome if a benzodiazepine is included in the anaesthetic regimen. Administration of fentanyl/fluanisone (0.3 ml/kg i.m. or 0.65 ml/kg i.p.) and either midazolam (2.5 mg/kg) or diazepam (2.5 mg/kg) produces medium surgical anaesthesia, excellent muscle relaxation and only a mild degree of respiratory depression (Flecknell and Mitchell, 1984). The commercial preparation of fentanyl/fluanisone ("Hypnorm") and midazolam (Hypnovel, Roche) can be combined and given as a single injection if prediluted with water. A mixture of one part fentanyl/fluanisone plus one part water and one part midazolam plus one part water is stable for at least 2 months, and can be given as a single intraperitoneal injection to produce good surgical anaesthesia in rats (2.7 ml/kg). If diazepam is administered, this must be given as a separate i.p. injection.

Fentanyl/droperidol: fentanyl/droperidol (Innovar Vet) differs markedly from fentanyl/fluanisone in its effects in the rat. Administration (0.4 ml/kg i.m.) produces profound analgesia, but this is associated with an even greater degree of muscle rigidity. Some animals adopt very abnormal postures, and at higher dose rates severe respiratory depression is produced. Attempts to combine this commercial preparation with benzodiazepines have not proven successful, and it is recommended that if this drug combination is used it is used only to provide restraint for superficial surgery.

Etorphine/methotrimeprazine: etorphine/methotrimeprazine (Immobilon SA) (0.5 ml/kg i.m.) produces profound analgesia, a poor degree of muscle relaxation, and profound respiratory depression. If administered with midazolam (0.25 ml/kg + 2.5 mg/kg i.m.), surgical anaesthesia is produced, together with excellent muscle relaxation, but moderate or severe respiratory depression. The duration of anaesthesia (45–60 min) is longer than that produced by fentanyl/fluanisone/midazolam.

Other Agents:

Metomidate and etomidate: both of these compounds produce hypnosis (sleep) but have no analgesic action when administered alone. Etomidate may be combined with a potent opioid such as fentanyl to produce surgical anaesthesia (DeWildt et al., 1983).

Chloral hydrate: chloral hydrate (400 mg/kg i.p.) produces 45–60 minutes light surgical anaesthesia following intraperitoneal injection. However, the depth of anaesthesia produced varies considerably between different strains of rat, and in some strains a depth of anaesthesia sufficient to allow major surgery is attained. Chloral hydrate has been reported to be particularly useful when studying CNS function, as it may have less depressant effects on neuronal function than other anaesthetics. Unfortunately, use of chloral hydrate is associated with a number of undesirable side effects. A high incidence of fatal paralytic ileus has been reported following its use in the rat (Fleischman et al., 1977). Chloral hydrate has also been reported to produce gastric ulceration after administration of normal anaesthetic dose rates (Ogino et al., 1990). Because of the risk of undesirable side effects and the very variable depth of anaesthesia that is produced, alternative anaesthetic regimens should be used whenever possible. If chloral hydrate is to be used, then a dilute solution (36 mg/ml) should be administered, as it has been suggested that this may result in a reduced incidence of some side effects (Fleischman et al., 1977).

Chloral hydrate/magnesium sulphate/pentobarbitone ("Equithesin") (3 ml/kg i.p. of commercial preparation): this mixture of two anaesthetic agents and a compound which is more usually administered as a euthanasia agent (magnesium sulphate) in large animals can be used to produce light surgical anaesthesia in the rat. In some strains, a depth of anaesthesia sufficient for major surgery is produced. The problems associated with the use of this mixture relate primarily to the inclusion of chloral hydrate. Paralytic ileus can be produced in the postanaesthetic period, and this can result in the deaths of large numbers of animals.

Tribromoethanol: tribromoethanol (125–300 mg/kg i.p.) has been reported to produce medium planes of surgical anaesthesia in the rat (Green, 1987). If it is to be used for recovery anaesthesia, it is essential to prepare a fresh solution on each occasion that it is to be administered, since decomposition of the material can result in peritonitis, gut disorders, and death of the animal following its use (Nicol et al., 1965; Tarin and Sturdee, 1972). Administration of tribromoethanol on a subsequent occasion can result in peritonitis and death, even when a freshly prepared solution is used (Norris and Turner, 1983). Because of the risk of adverse effects following anaesthesia, alternative anaesthetics should be used whenever possible.

Pentobarbitone: this barbiturate has been one of the most widely used injectable anaesthetics for laboratory rats. It has the advantage that a single i.p. injection (40 mg/kg) can be given to produce light surgical anaesthesia. It has a narrow safety margin, and until an appropriate dose for a specific strain, sex and age of rat has been established, an unacceptably high mortality rate is likely to be encountered if surgical anaesthesia is produced. Pentobarbitone produces severe cardiovascular and respiratory depression. Recovery is prolonged and no specific antagonist is available.

Urethane: see 10.1.

Chloralose: see 10.1.

8.4. Recommended Techniques

As discussed earlier, the choice of a particular anaesthetic regimen will be influenced by the overall objectives of the experiment. If not contraindicated because of interactions with the experimental protocol, the following anaesthetic regimens should be used since they provide safe and effective surgical anaesthesia. Anaesthesia for very prolonged periods raises a number of special considerations, and these are discussed separately in Section 10.

8.4.1. Short-term Anaesthesia (1–10 minutes)

Very brief periods of general anaesthesia can be provided either by use of a volatile anaesthetic, by intravenous injection of a short-acting injectable anaesthetic, or by the use of a reversible anaesthetic combination. Provided that the apparatus is available, and the nature of the procedure does not prevent their use, then there can be little doubt that use of a volatile anaesthetic represents the most simple and effective means of producing a short period of anaesthesia in the rat. If a volatile agent cannot be used, then intravenous administration of propofol or alphaxalone/alphadolone is recommended as the most suitable alternative.

8.4.2. Medium-term Anaesthesia (10–60 minutes)

Moderate periods of general anaesthesia can be provided either by use of a volatile anaesthetic, by intraperitoneal or intramuscular administration of injectable anaesthetics or by continuous intravenous infusion of a short-acting anaesthetic. The most useful combinations for achieving medium duration surgical anaesthesia are either fentanyl/fluanisone ("Hypnorm") combined with midazolam or ketamine in combination with either xylazine or medetomidine. The other anaesthetics and anaesthetic combinations described above are generally less satisfactory and have a narrower margin of safety. If the use of volatile anaesthetics is practicable, then research workers are strongly recommended to consider their use since they allow easy alteration of depth of anaesthesia and rapid recovery.

8.4.3. Combined Injectable/Inhalational Regimens

Several of the disadvantages of producing anaesthesia by the exclusive use of an inhalational agent or an injectable agent can be overcome by combining these techniques. For example, if an induction chamber is not available, then administration of a sedative or sedative/analgesic combination (Table 7) can prevent any struggling or distress caused by administration of anaesthetic by face mask. Administration of an injectable anaesthetic, to induce anaesthesia, followed by maintenance with a volatile agent, also avoids problems of restraint during induction.

Some injectable anaesthetic combinations may fail to provide sufficient analgesia for major surgery (e.g. laparotomy), and the addition of an inhalational agent at low concentration (e.g. 0.25–0.5% halothane) may be a safer and more convenient technique than trying to "top up" with additional injectable agent. Similarly, when using an inhalational agent as the major component of an anaesthetic regimen, the concentration that is required to produce surgical anaesthesia can be reduced by administering a potent analgesic (e.g. fentanyl). During prolonged procedures, this supplementation can be given intermittently during periods of major surgical stimulation. The aim of this balanced anaesthetic regimen is to minimise the interference to the animal's physiology caused by the drugs used, and so enable recovery to be as smooth and as rapid as possible.

8.4.4. *Reversal of Injectable Anaesthetic Regimens*

An alternative approach to solving some of the problems associated with the prolonged effects and varying responses to the i.p. or i.m. injection of anaesthetics is to use an anaesthetic regimen which can be reversed using specific antagonist drugs. Several antagonists are available commercially and these can partially reverse some anaesthetic combinations.

The antagonists currently available in the USA and UK are listed in Table 8. Reversal, or partial reversal, of anaesthesia can have several major benefits. In reversing the sedation and reducing sleep time, animals become active more rapidly, and will no longer be at risk due

Table 8. Anaesthetic antagonists for use in rats.

Compound	Anaesthetic regime	Dose rate	Comments
Doxapram	All anaesthetics	5–10 mg/kg i.m., i.v., i.p.	General respiratory stimulant
Atipamezole	Any regime using xylazine or medetomidine	1 mg/kg i.m., i.p., s.c., i.v.	Highly specific antagonist, available in USA 1992
Yohimbine	Any regime using xylazine or medetomidine	2.1 mg/kg i.p.	Relatively non-specific antagonist not recommended
Naloxone	Any regime using μ opioids (e.g. fentanyl)	0.01–0.1 mg/kg i.v., i.m., i.p.	Reverses analgesia as well as respiratory depression
Nalbuphine	Any regime using μ opioids	1–2 mg/kg i.p. or s.c.	Almost as rapid acting as naloxone maintains postop. analgesia
Buprenorphine	Any regime using μ opioids	0.05 mg/kg i.p. or s.c.	Slower onset than naloxone and nalbuphine, but longer duration of action

to hypothermia. Many anaesthetic drugs depress respiration, and this depression can be promptly and effectively reversed with antagonist drugs. An additional advantage can be obtained when reversing an agonist opioid such as fentanyl with a mixed agonist/antagonist opioid such as nalbuphine or buprenorphine. Under these circumstances the respiratory depression and muscle rigidity produced by fentanyl are reversed, as is its analgesic effect, but analgesia persists because of the agonist/antagonist's mixed actions at opioid receptors. In simple terms, the undesirable effects of the anaesthetic are reversed, but postoperative analgesia persists for several hours (Flecknell *et al.*, 1989).

8.5. Assessment of Depth of Anaesthesia

Numerous attempts have been made to translate Guedel's classical description of the stages of anaesthesia with ether in man (Guedel, 1937) into a universal scheme applicable to all anaesthetics in all animals. Such attempts provide schemes that appear useful, but which have very little practical value. Rather than attempt to classify depth of anaesthesia into a large number of stages and planes, a more pragmatic approach has been adopted in this section. Light anaesthesia is defined as the point at which the animal loses its righting reflex and is immobilised, but responds markedly to painful stimuli. Light surgical anaesthesia indicates a deeper level of anaesthesia, allowing minor superficial surgical procedures such as a skin biopsy to be undertaken. Onset of medium surgical anaesthesia enables procedures such as laparotomy to be carried out without the animal moving in response to surgical stimuli. Deep surgical anaesthesia may be needed for some major procedures or when operating on particularly sensitive structures. For example, scraping the periosteum from the surface of the cranium appears to require deep surgical anaesthesia to prevent the animal from moving.

Monitoring of depth of anaesthesia can be considered simply as ensuring that a degree of analgesia has been produced which is sufficient to prevent the animal perceiving pain during surgery, and that sufficient loss of consciousness has occurred to prevent distress caused by restraint. Since animals cannot report their experiences during anaesthesia, we can only use the presence or absence of certain reflex responses to judge the depth of anaesthesia. The most useful response is the pedal withdrawal reflex. The animal's hind limb is extended and the skin between the digits pinched firmly between the operator's fingernails. If the animal attempts to withdraw its limb, then it is not sufficiently anaesthetised to allow major surgery to be undertaken. If there is no response, then medium or deep surgical anaesthesia has been produced. The response to pinching the tail can also be assessed, but this reflex usually disappears when only light surgical anaesthesia has been produced.

9. PATIENT CARE

9.1. Preoperative Preparation

Many of the problems that may arise during anaesthesia can be minimised by ensuring that the patient is both in overt good health and free from subclinical disease. Whenever possible animals of defined health status should be obtained, so that the occurrence of conditions such as respiratory infections can be minimised.

In some instances the health of the patient will be uncertain, or may have been altered by previous experimental procedures. In any event, non-infectious disease may be present so that a clinical examination should be carried out on each animal. The scope of such an examination will depend to some extent on the type of procedure that is to be undertaken, but some attempt should always be made to assess the adequacy of cardiovascular and respiratory function. It should be ensured that food and water intake are normal, and that the rat is free from signs of clinical disease. This assessment can only be meaningful if the animal has been allowed to acclimatise to the particular husbandry regimen in the research institute. Acclimatisation times vary, but in general a 1–2 week period should be allowed prior to performing anaesthesia and surgery.

Preanaesthetic fasting of rats is unnecessary since vomiting during induction does not occur in this species. Fasting is only likely to be necessary if upper gastrointestinal (GI) tract surgery is to be undertaken. Even then, since coprophagy is often not prevented, removing the animal's food may not result in an empty stomach. We have experienced no problems in allowing free access to both food and water until immediately prior to anaesthesia in rats required to undergo a wide variety of surgical procedures.

9.2. Intraoperative Care

Three essential areas need to be considered when planning the intraoperative care of an animal. These are the maintenance of respiratory function, the maintenance of the circulatory system, and most importantly in small animals such as rats, the maintenance of body temperature.

9.2.1. Respiratory System

Virtually all anaesthetics cause some depression in respiration, and this will result in both an elevation of blood carbon dioxide tensions (hypercapnia) and a reduction in blood oxygen tensions (hypoxia). The effects of hypercapnia, whilst undesirable, are made worse by concurrent hypoxia. Providing the animal with oxygen intraoperatively usually prevents the development of hypoxia, and can be life saving. Provision of oxygen is simple and inexpensive, requiring only an oxygen cylinder, a regulator/flowmeter and a face mask.

9.2.2. Monitoring Respiration

It is difficult to provide adequate monitoring of respiratory function in rats, since many respiratory monitors do not function effectively in these small animals. Some commercially available instruments which use a thermistor to detect respiration can be used successfully in larger rats (>200 g) (e.g. IMP respiration monitor, see Equipment Index). If possible, the instrument should be obtained for a trial period to assess its reliability. If suitable electronic equipment is unavailable, the anaesthetist should assess the adequacy of respiratory function by recording the respiratory rate, and noting the pattern and depth of respiration, and the colour of the mucus membranes. In albino animals, the colour of the ears, muzzle and footpads gives a reasonable indication of the degree of hypoxia. Development of a bluish coloration indicates onset of severe hypoxia which will usually require corrective therapy (e.g. administration of oxygen and assisted ventilation).

Although it is helpful to record preanaesthetic respiratory rates, these are often abnormally high, because of the disturbance caused by handling and restraint. The respiratory rate in normal (undisturbed) rats is usually between 70 and 90 breaths per minute, and in general a fall of up to 50% from this value can be considered acceptable. Progressive reduction in respiratory rate usually indicates a serious problem, and some corrective action should be taken. It is important that an attempt is made to assess why respiratory depression or cyanosis is occurring. If this has arisen because of blockage of the oropharynx or upper respiratory tract with mucus, blood, or other material, or from abnormal positioning of the head and neck, then the airway should be cleared using a suction device. A simple but effective suction apparatus can be constructed by connecting a 16G catheter to a 20 ml syringe. The positioning of the head and neck should be checked, and any twisting or flexion corrected. The chest or neck may inadvertently be compressed by the surgeon or by surgical instruments, and this can be overlooked, particularly if the rat is covered with a surgical drape. If oxygen is not already being administered, commencing oxygen therapy will be beneficial. If surgery has not commenced, consider giving an antagonist to reverse the anaesthetic, or a general respiratory stimulant. If surgery has started, assist ventilation by manually squeezing the chest between your finger and thumb, at a rate of approximately 60 breaths/min, and try to complete the surgical procedure as rapidly as possible. If you have intubated the animal (see 10.6), then ventilation can be assisted easily, either manually or by using a mechanical ventilator.

9.2.3. Cardiovascular System

Blood loss, hypothermia, and anaesthetic overdose can all cause cardiovascular failure or death. Blood loss is a common cause of fatality in rats, primarily because the degree of haemorrhage relative to the animal's blood volume is not appreciated until irreversible hypovolaemia has occurred. A 200 g rat has a circulating blood volume of approximately 15 ml. Rapid loss of 3–4 ml of blood will be sufficient to cause signs of cardiovascular failure ("shock"). Careful attention to surgical technique, and consideration of use of a blood donor if haemorrhage occurs midway through a complicated procedure, can help to minimise this problem. Blood transfusions are rarely considered when carrying out surgery in rats, but suitable donors are often readily available. If an inbred strain is being used, then cross-matching of donor and recipient is unnecessary. Even when working with outbred rats, adverse reactions to transfusions are rare. In an emergency, blood can be collected without use of any anticoagulant and administered within about a minute. It is preferable to collect blood in acid citrate dextrose solution (three parts blood to one part ACD see 59), as it can then be stored for several days at 4°C until required.

Monitoring of the cardiovascular system can provide an early indication of impending cardiovascular failure, but such monitoring is rarely used during anaesthesia of rodents. If a long period of anaesthesia is planned, or if unexpected mortality is encountered, then monitoring of arterial blood pressure and of heart rate should be considered. Non-invasive measurement of arterial blood pressure is possible using a tail cuff technique (see 40.1), or alternatively the carotid artery can be cannulated and direct measurements made using a suitable pressure transducer (see 22.3 and 40.2). Monitoring of the ECG is feasible, but many of the monitors designed for use in man are unable to detect such low amplitude signals. An additional problem is that many of these monitors are unable to register heart rates above 250 or 300 beats/min. One monitor which is capable of monitoring heart rate and the ECG in rats and other small rodents is the EC-60 available from Silogic Ltd (see Equipment Index) (Fig. 67).

9.2.4. Hypothermia

All anaesthetics affect thermoregulation, and an animal's body temperature will fall during anaesthesia unless measures are taken to prevent this. Small animals are particularly susceptible to this problem, since they have a large surface area available for heat loss, relative to their small body mass. It is important to try to prevent heat loss occurring, rather than to try to revive a severely chilled animal. Clearly, careful pre- and intraoperative management can minimise any fall in body temperature. All anaesthetised rats will require some additional heating, and should be insulated to minimise heat loss. Effective insulation can be provided either by wrapping the animal in cotton wool, followed by an outer wrapping of aluminium foil, or by using the "bubble packing" which frequently forms part of the packaging of laboratory equipment.

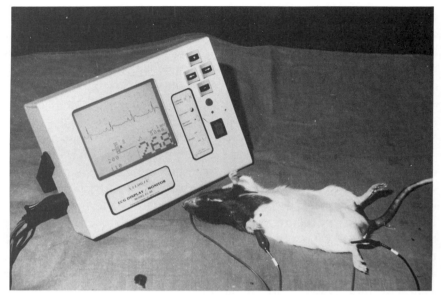

Fig. 67. Monitoring the electrocardiogram (ECG) using an electronic monitor (EC-60, Silogic Design).

Fig. 68. Use of "bubble packing" to minimise heat loss. A window can be cut to expose the operative field.

After wrapping the animal in an insulating layer of material, a "window" can be cut to expose the operative field (Fig. 68). Additional heating can be provided by heat lamps and heating blankets, but care must be taken not to burn the animal. A thermometer placed adjacent to the animal will indicate whether excessive heat is being applied — the temperature should not exceed 40–42°C. It is preferable to use a thermostatically controlled heating blanket, regulated by the animal's body temperature using a rectal probe (e.g. Harvard homeothermic system, see Equipment Index). If such a unit is not available, a simple heating pad or overhead lamp can be used, and provided that the animal's rectal temperature is monitored (Fig. 69), these can be switched on and off manually as required.

The importance of this monitoring cannot be overemphasised. Hypothermia is undoubtedly the commonest cause of mortality during anaesthesia of rats. It is also important to ensure that measures to prevent hypothermia are continued throughout the recovery period (see below).

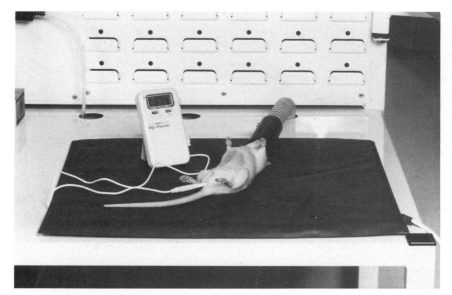

Fig. 69. Maintenance of body temperature using a simple heating pad. Body temperature is monitored continually using an electronic thermometer. The inexpensive electronic thermometer enables upper and lower temperature alarms to be preset (see Equipment Index).

10. NON-RECOVERY AND LONG-TERM ANAESTHESIA

Many experimental protocols require the induction and maintenance of terminal anaesthesia, or require periods of anaesthesia in excess of that provided by a single dose of the commonly available injectable anaesthetic combinations. There is little difference between recovery and non-recovery anaesthesia in terms of patient care and management, although the approach used for procedures such as vascular cannulation may differ. Similarly, tracheostomy rather than oral endotracheal intubation is more likely to be used to establish and maintain an airway if the animal is not to be allowed to recover. Some anaesthetic agents are best reserved solely for non-recovery experiments.

When developing regimens for long-term anaesthesia, three options can be considered. Either a single dose of a long-acting anaesthetic such as urethane or chloralose can be administered, or shorter acting agents can be administered either by intermittent injection or by continuous infusion. A third option is to induce and maintain anaesthesia using a volatile agent.

10.1. Long-acting Anaesthetics

10.1.1. Inactin

Inactin is a barbiturate which has been reported to provide surgical anaesthesia following i.p. injection in the rat (80 mg/kg) (Buelke-Sam et al., 1978). In the author's experience the effects of this agent vary considerably depending upon the strain of rat. In some instances, long-lasting stable anaesthesia is produced, but in other strains anaesthesia is too light to allow surgery to be undertaken, and a small increase in dose rate causes death due to respiratory failure. As with many other agents, it is important to assess the efficacy and suitability of this anaesthetic in the particular strain of rat that is to be used.

10.1.2. Urethane

Urethane (1000 mg/kg i.p.) produces stable, long-lasting anaesthesia in rats, with maintenance of good cardiovascular reflexes (Buelke-Sam *et al.*, 1978; DeWildt *et al.*, 1983). Unfortunately it is a carcinogen (Field and Lang, 1988) and so should only be used when no suitable alternative is available. If urethane is used, it should be prepared and administered under controlled conditions, observing the safe handling precautions appropriate when dealing with carcinogens. It should be noted that urethane is highly irritant and can cause peritonitis, and so should not be used for recovery procedures (Van der Meer *et al.*, 1975).

10.1.3. Chloralose

Chloralose (55 mg/kg i.p.) produces light anaesthesia in the rat. If higher dose rates are used (65 mg/kg i.p.), medium planes of surgical anaesthesia can be produced in some strains of rat. Anaesthesia lasts for up to 8–10 hours. Recovery is prolonged and associated with extended periods of involuntary excitement, so that it is recommended only for use in non-recovery surgery. Induction can also be associated with a prolonged period of involuntary excitement, so it is preferable first to induce anaesthesia with a volatile agent, then administer the chloralose, and finally discontinue administration of the volatile anaesthetic.

10.1.4. Chloralose/Urethane

A combination of chloralose (80 mg/kg i.p.) and urethane (400 mg/kg i.p.) has been reported to produce long-term anaesthesia in the rat. The combination produces a smoother and more rapid induction of anaesthesia than does chloralose alone.

10.2. Intermittent Administration of Short-acting Agents

Prolonged anaesthesia can be produced by the repeated injection of one or more anaesthetics. Although this approach can be used successfully, inevitably the plane of anaesthesia will vary markedly. In most instances, a top-up dose is administered in response to signs of emergence from anaesthesia, such as spontaneous movement. This approach can lead to animals experiencing pain because of maintenance of inadequate levels of anaesthesia. It is preferable to administer the anaesthetic by continuous infusion so that a stable plane of anaesthesia can be maintained.

10.3. Continuous Intravenous Infusion

Several long-term anaesthetic regimens have been established using continuous i.v. infusion of relatively short-acting anaesthetics such as alphaxalone/alphadolone, propofol, fentanyl/midazolam and etomidate/fentanyl. The use of a short-acting anaesthetic enables the plane of anaesthesia to be adjusted rapidly, and recovery is likely to be rapid if the agent is non-cumulative in its effects. For this reason agents such as propofol (0.6–1.5 mg/kg/min i.v.) and alphaxalone/alphadolone (0.2–0.7 mg/kg/min i.v.) are preferred rather than the short-acting barbiturates such as thiopentone. These infusion rates should only be taken as a general guide, given the considerable variation between different strains of rat. After establishing anaesthesia, and selecting an infusion rate, the animal should be carefully monitored to assess any changes to the depth of anaesthesia. Appropriate changes can then be made to the infusion rate—usually a change in steps of ±20% is required initially. After 30 minutes or so the animal usually stabilises. After 1 or 2 hours, infusion rates can often be lowered, especially if all surgical procedures have been completed.

10.4. Inhalational Agents

All of the inhalational anaesthetics discussed earlier can be used to provide prolonged periods of anaesthesia. If the animal is to recover from anaesthesia, then the concentration of the agent should be reduced during the final stages of wound and skin suturing to try to reduce the time taken for recovery.

10.5. Management of Long-term Anaesthesia

When anaesthetising animals for prolonged periods, factors such as maintenance of body temperature, respiratory function and cardiovascular stability are of considerable importance. In most instances it is advisable to intubate the rat or carry out a tracheostomy so that intermittent positive pressure ventilation can be carried out. Even if the rat is not to be ventilated, tracheostomy or intubation (see below) are still advisable since it enables a clear airway to be maintained and allows ventilation to be assisted without delay should this prove necessary.

A secure venous line should be established for infusion of fluids to provide circulatory support and to enable administration of anaesthetic and analgesic drugs. In addition, drugs to provide circulatory and respiratory stimulation can be given rapidly if an emergency arises. Provided that blood loss does not exceed 0.7 ml/100 g body weight, then a balanced electrolyte solution (e.g. Hartmann's solution) should be infused at a rate of approximately 2 ml/100 g/h.

Aside from instituting monitoring of vital signs, it is also useful to monitor the continued function of the anaesthetic delivery system during prolonged anaesthesia. Infusion pumps should be fitted with occlusion alarms, and alarms to alert the operator that the syringe or drug reservoir is empty. Anaesthetic machines should include an alarm to warn of failure of the oxygen supply. It is also important to provide sufficient personnel to manage the experiment. A research worker undertaking long-term anaesthesia and carrying out the experiment will become fatigued and need regular rest periods. If assistance is not available, then it may not be acceptable to carry out the procedure.

10.6. Endotracheal Intubation

Endotracheal intubation of rats can be achieved using a variety of techniques. One approach which has been found most useful by the authors is to construct a small laryngoscope blade (Costa *et al.*, 1986) and pass an endotracheal tube whilst viewing the larynx (Fig. 70). A 14–18 gauge catheter should be used as the endotracheal tube. It is advisable to push a small (3 mm) length of rubber tubing of appropriate diameter onto the catheter approximately 0.5–1.0 cm from the tip to act as a seal against the larynx, and to prevent the tube being inserted too far and inadvertent intubation of one bronchus.

10.7. Assisted Ventilation

A number of commercially produced rodent ventilators are available (e.g. Harvard rodent ventilator, see Equipment Index; Fig. 71), and these can be used to provide a safe and effective ventilation provided

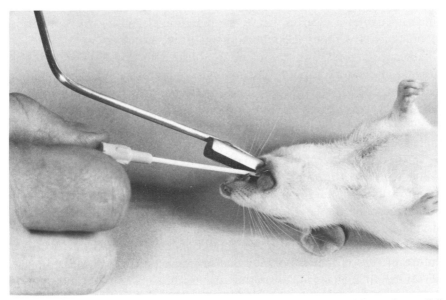

Fig. 70. Endotracheal intubation using a purpose-made laryngoscope and "over-the-needle" catheter.

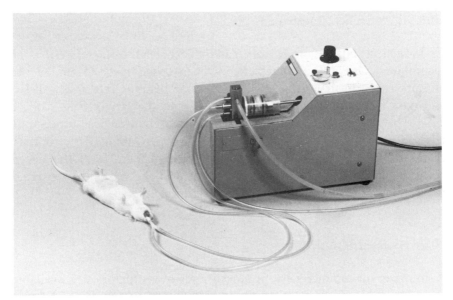

Fig. 71. Volume-cycled rodent ventilator (Harvard Instruments Ltd, see Equipment Index).

that a few simple precautions are observed. Other ventilators designed for human or veterinary use almost invariably prove incapable of delivering a sufficiently small tidal volume (tidal volume is the volume of gas drawn in and out of the lungs with each breath). It is possible to construct an anaesthetic circuit to allow these ventilators to be used, and if no alternative exists then possible modifications should be discussed with an experienced anaesthetist.

Commercially produced rodent ventilators perpetuate the commonly held misconception that ventilation rates for rats should be high—one instrument offers rates of up to 200 breaths/min. In normal circumstances a rate of 50–60 breaths/min is recommended, with a tidal volume of 1.5 ml/100 g. It is important to ensure that the animal's lungs are not subjected to excessive pressures by the ventilator. The pressure indicators on some commercial ventilators are misleading when the instrument is connected to a rat using relatively fine diameter tubing (e.g. 0.5–1.0 cm). If possible, when commencing a series of experiments with a new ventilator, measure the airway pressures produced by connecting a pressure transducer to a side arm as close as possible to the endotracheal tube. In general, pressures of 5–15 mmHg should be sufficient to ensure good lung inflation.

The type of circuit used will depend upon the ventilator design. If the ventilator has a valve-controlled outlet to the lungs, and a valve-controlled expiratory inlet, then connection to the animal can be by means of two tubes and a simple "Y" connector attached to the endotracheal tube (Fig. 71). If the ventilator acts by delivering a jet of fresh gas, then a T-piece circuit can be used. It is important to ensure that equipment dead space is minimised, and that some pressure relief system is provided if the ventilator is not equipped with this safety device. This is particularly important when the inspired gas is provided from an anaesthetic machine, since the pressures which can be delivered can be sufficient to damage the animal's lungs. The easiest means of introducing a pressure relief system is to include a standard anaesthetic rebreathing bag and pop-off valve in the connection between the anaesthetic machine and the ventilator.

11. ANAESTHESIA OF HIGH-RISK ANIMALS

In some circumstances it is necessary to anaesthetise animals which have an increased risk of anaesthetic-related death. Animals with spontaneous or experimentally produced hypertension, diabetes or other endocrine imbalance, neonatal or geriatric animals and pregnant animals all have an increased risk of death during a period of anaesthesia. In all instances, survival is increased by meticulous attention to intra- and postoperative care, especially to measures to prevent hypothermia and hypoxia. The response to injectable anaesthetics varies considerably in some of these groups of high-risk animals, for example amongst neonates and aged rats. For this reason, it is recommended that volatile anaesthetics be used whenever possible, so that the anaesthetic depth can be varied rapidly depending upon the response of the individual animal. It should be noted that the potency of volatile anaesthetics varies with the age of the animal. Neonatal animals require a higher concentration of anaesthetic, aged animals require a lower concentration. When using halothane in young adult rats, a concentration of approximately 2% is usually required to maintain surgical anaesthesia. Neonatal rats may require 2–3% and aged rats 0.8–1.5% If problems are encountered, consider using a low concentration of volatile anaesthetic to produce unconsciousness, and infiltrate the surgical field with local anaesthetic to provide surgical analgesia.

12. POSTOPERATIVE CARE

Many of the monitoring procedures and routine therapeutic regimens described earlier should be continued in the immediate postoperative period. There is often no clear demarcation between the intra- and postoperative care regimens, and it is usually preferable for the same staff to assume responsibility for both periods of patient care. This helps to ensure that optimal attention is given to the animal, and that all the measures needed to ensure its comfort and well-being are undertaken promptly and effectively.

Since all animals will require some degree of special attention in the postoperative period, it is preferable to provide a separate recovery area. This not only enables more appropriate environmental conditions to be maintained but also encourages individual attention and special nursing. Maintenance of effective postoperative analgesia is of fundamental importance, and this is dealt with in detail below.

The recovery area should be stocked with a range of emergency drugs and equipment. Additional apparatus such as monitoring equipment may be needed after major surgery and/or prolonged anaesthesia.

12.1. Warmth and Comfort

The provision of supplemental heating to prevent hypothermia is of major importance. This can often be achieved most effectively by the use of animal incubators (Fig. 72) (see Equipment Index). A temperature of 25–30°C is needed for adult animals, 35–37°C for neonates. If an incubator is unavailable, heating pads and lamps should be provided.

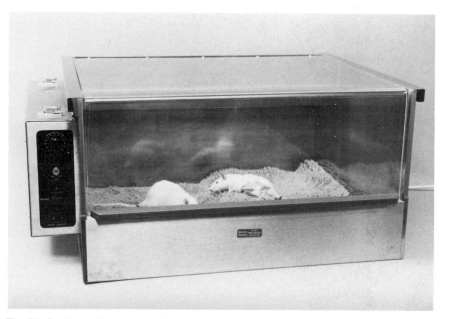

Fig. 72. Incubator for postanaesthetic recovery.

Care must be taken not to overheat the animal, and a thermometer should be placed next to the animal to record its surface temperature.

The bedding used should be both comfortable and provide effective insulation. Rats should never be allowed to recover from anaesthesia on sawdust bedding, since this will stick to the animal's eyes, nose and mouth. At the least, large quantities of tissue paper should be provided, and it is preferable to use towelling or specially produced bedding material (e.g. "Vetbed", see Equipment Index, Fig. 73). Body temperature should be monitored carefully postoperatively, to ensure that the above measures are effective.

12.2. Respiratory Depression

Respiratory depression commonly develops postoperatively, and is particularly dangerous since it frequently goes unnoticed until severe hypercapnia and hypoxia have developed. It is obviously preferable to continue to monitor respiration, for example by using a thermocouple

Fig. 73. Bedding material for use during anaesthesia and in postanaesthetic period.

positioned close to the animal's nose. If respiratory depression develops, it should be treated by the administration of a respiratory stimulant such as doxapram (5–10 mg/kg i.v. or i.p.). If depression recurs, the doxapram should be administered repeatedly, at approximately 10–15 minute intervals.

12.3. Fluid Therapy

Excessive losses of body fluid often occur intraoperatively and in addition the animal may not drink for the first 12–24 hours postoperatively. Fluid should be given to replace the estimated losses, and also to provide sufficient water intake for the next 12–24 hour period. Fluid requirements are approximately 40–80 ml/kg/24 h for most animals, and this is best given by the oral route if the animal is fully conscious. In animals which are unwilling or unable to accept oral fluids, subcutaneous or intraperitoneal administration of dextrose–saline (4% dextrose, 0.18% saline) or saline (0.9%) is a quick and easy technique (Fig. 74). If severe dehydration is present, as judged by loss of skin tone and "tenting" of

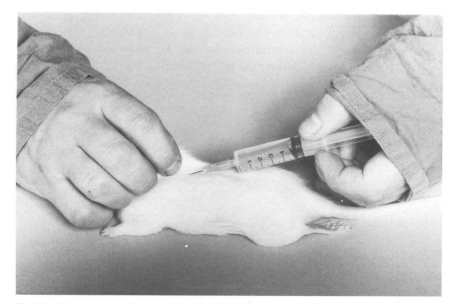

Fig. 74. Subcutaneous administration of fluids to help prevent postoperative dehydration.

the skin when it is pinched, then fluids must be administered intravenously. Monitoring of body weight pre- and postoperatively can provide a good indication of the adequacy of fluid intake. The urine and faecal output of the animal should be recorded, and any abnormalities investigated. Reduced urine output may be due to dehydration, urinary tract injury, or because the animal is suffering pain.

Veterinary advice should also be sought if the animal fails to defaecate. Whilst this may be due simply to an absence of faeces because of preoperative fasting, it may also be caused by paralytic ileus, due to excessive handling of the bowel at laparotomy.

12.4. General Care

The value of personal contact and postoperative nursing attention differs between animal species. Companion animals such as dogs and cats respond well to personal contact, but this may have adverse, stressful effect on rats. Rats often prefer to be left in a warm environment with subdued lighting, and be disturbed as little as possible. Nevertheless, an attempt should be made to keep the animal clean and dry, since this will be both more comfortable for the animal and will also encourage closer attention from the nursing staff. A bright light source should be readily available to enable the animal to be examined. All clinical observations and drugs administered must be recorded on a patient record card that should ideally be left attached to the incubator or cage for rapid reference.

13. ASSESSMENT AND ALLEVIATION OF POSTOPERATIVE PAIN

13.1. Animal Pain

Pain in man has been defined as: "An unpleasant sensory and emotional experience associated with actual or potential tissue damage or described in terms of such damage" (Definition of the Taxonomy Committee of the International Association of the Study of Pain).

It is reasonable to assume that the sensory experiences of pain are present in animals. The mechanisms for nociception in animals have been extensively studied (see, for example, reviews in Kitchell *et al.*, 1983) and found to closely resemble those present in man. The central pathways that are presumed necessary for the emotional component of pain are also present. Nevertheless, the demonstration of equivalent anatomical structures and physiological processes does not provide conclusive evidence that the sensory/emotional experience of pain in animals is identical to that experienced by humans. It is possible that the relative significance, magnitude and duration of pain in response to particular types of injury or tissue damage may all vary in animals. Although it may be tempting to "downgrade" the significance of pain in animals and use this to account for their apparently rapid recovery from some surgical procedures, such reasoning often leads to a failure to provide effective pain relief. Until a more detailed knowledge of the nature of pain in animals has been obtained, it should be assumed that animals experience pain, that this experience is unpleasant and would normally be avoided, and consequently that it should be alleviated whenever possible.

Pain, of course, is of value when it serves as a warning of damage to an animal. Pain arising from injured tissues can serve to immobilise the affected area and so help to prevent further injury. Immobilisation of an injured limb may also promote healing. Nevertheless, pain is also harmful since the immobility and muscular spasm it produces can cause weakness, muscle wasting and if prolonged this may produce a permanent abnormality of gait. Thoracic and abdominal pain may reduce ventilation and cause hypoxia and hypercapnia. Prolonged immobility may result in pressure sores and deep vein thrombosis. Pain also appears to result in reduced food and water intake (Flecknell and Liles, 1991, 1992; Morton and Griffiths, 1985). Pain in man has been shown to prolong the metabolic response to surgery (Kehlet, 1978), and to increase the requirement for hospital care following operative procedures (Alexander and Hill, 1987).

It is frequently stated that analgesics should be withheld, or provided in minimal quantities, so that the protective function of pain is maintained. Provided that effective measures are taken to immobilise or protect damaged tissues, and that competent surgical techniques are employed, the abolition of pain rarely gives rise to problems. If it is believed that excessive activity may occur following the provision of effective pain control, then a sedative or tranquilliser should be administered in addition to analgesics.

13.2. Methods of Assessment of Pain

Pain must be assessed if it is to be diagnosed and treated, and in man a variety of schemes has been devised to achieve this (Alexander and Hill, 1987). The majority of these schemes rely on the patient responding verbally, or by choosing descriptive words, or by completing a visual analogue score chart. Objective or indirect measurement of pain in man has proven difficult, but attempts have been made to use parameters such as pattern and depth of respiration, certain facial expressions and behaviour, and autonomic or biochemical changes. Despite the considerable efforts which have been made to provide effective detection and relief of pain in man, numerous patients experience postoperative pain, and the management of chronic pain causes considerable problems (Smith, 1984).

Laboratory animal scientists face similar problems. Pain in animals is difficult to assess. We do not share a common language and inferences based on common experience cannot be made with much confidence. Nevertheless, two approaches to the recognition of pain are open to us. We can act on the statement "Pain is what the patient says hurts" (H.V. Beecher) and assume that if the animal could speak, it would communicate the presence of pain in association with conditions that would cause pain in man. Such an anthropomorphic approach has dangers, but offers a humane starting point for the management of postoperative pain. Although anthropomorphism may be acceptable, it must not be assumed that animals in pain will behave in the same manner as humans experiencing pain. The major problem of adopting an anthropomorphic approach to pain assessment is that it does not enable comparisons to be made of the efficacy of different analgesics. Neither does it allow an assessment of the dose rates, frequency of administration, or duration of therapy that are required.

A second approach relies upon using our experience to assess the clinical appearance of the rat and whether the animal is behaving in an abnormal manner (Morton and Griffiths, 1985; AVTRW, 1989). The behaviour of different individual rats, and of rats of different strains, varies considerably. Any method of assessment will require that the clinician is thoroughly familiar with the normal behaviour of the strain of rat, and preferably that he is familiar with the normal behaviour of the individual animal.

After undergoing a surgical procedure which would cause significant postoperative pain in man, most rats will modify their behaviour. Frequently, they reduce their overall level of activity, and will often remain immobile in one area of their cage or pen for prolonged periods. Some rats appear to neglect normal grooming behaviour, and in consequence their coat may become unkempt and dirty. Porphyrin staining of the periorbital and nasal region may be present. Porphyrin appears as a black or reddish-brown discharge, and is a non-specific response to stress. If the discharge is wiped with a damp swab, it will stain the swab a bright red colour. If attempts are made to force the animal to move, it will usually be very reluctant to do so, and may respond in an uncharacteristically aggressive manner. Rats may also vocalise, often with an altered pitch. The animal's posture and gait when it does move will often be noticeably abnormal. For example, following abdominal surgery, a rat may adopt a hunched-up posture, and may show an alteration in hind limb movement to avoid tension on the surgical wound. Obviously, if the surgical wound is confined to one limb, this will not be used for normal movement and be held in an abnormal position to try to protect it from further trauma.

The behavioural changes which occur following surgery may be very subtle, and may be masked by the rat's response to being disturbed and observed. Rats are normally relatively very inactive during the light phase of their photoperiod, and active during the dark phase. Observing the rat in such a way that it is unaware of the presence of the observer, for example by using an observation panel, and using a red light source to allow viewing at night, may be necessary to obtain an accurate assessment of the animal's behaviour.

Aside from changes in behaviour, and alterations in response to handling, clinical examination of the animal may lead to a suspicion that pain may be present. Observation of the animal whilst it is undisturbed will allow the respiratory rate to be noted. This may be elevated if the animal is in pain, although the presence of thoracic pain may produce a depression of respiratory rate, accompanied by a reduction in the degree of movement of the thorax. If there is unusual rigidity of the musculature in any body area, or unusual response such as flinching or attempting to bite when an area is handled, this should be considered good evidence that pain is present.

Rats in pain will frequently reduce their food and water intake, and this will be reflected in an absence of faecal pellets and low urine output. It is helpful to monitor food and water intake for 2–3 days preoperatively, and over the first 2–3 days postoperatively, to determine whether food and water intake is depressed. Some depression will occur as a consequence of anaesthesia, but severe depression (>20%) should lead to a suspicion of the presence of pain. Recent studies suggest that water intake is least affected in normal animals by factors such as anaesthesia or administration of analgesics, but is reduced following surgery (Flecknell and Liles, 1991, 1992; Liles and Flecknell, 1992).

All of the criteria discussed above must be considered in conjunction with the nature of the surgical procedure which has been undertaken, and the previous behaviour of the animal. Although most of the criteria are somewhat subjective, and may be altered by factors other than pain, they nevertheless enable an overall clinical impression to be obtained. Some degree of quantification can be achieved by the adoption of a scoring system such as that described by Morton and Griffiths (1985). The technique requires the assessment of a range of clinical signs. Deviations from normality are scored on a scale of 1 to 3, a score of 1 indicating a mild degree of abnormality, a score of 3 indicating severe deviation from normal. In all, five groups of observations, appearance, food and water intake, body weight, behaviour and clinical signs, are made and scored. The overall score provides an indication of the state of the animal. The higher the score, the more likely it is that the animal is experiencing severe pain. A problem with this approach is that it utilises many clinical signs which are not specific indicators of pain, and which could be seen in an animal with a non-painful but serious illness. Despite this problem, an animal which has a high score is quite obviously abnormal, and should receive careful attention.

One difficulty of using this approach is that if administration of an analgesic reduces the total score then it is assumed that this is evidence that pain is present. Analgesics have a variety of effects other than pain relief. For example, buprenorphine, nalbuphine and butorphanol all cause an increase in activity in normal rats (Liles and Flecknell, 1992) and this can reverse the depression in activity which is frequently noted following surgical procedures. Other opioids have similar non-pain-related effects on behaviour and may also influence food and water intake, faecal and urinary output, and the interaction of the animal with others. Despite these problems, adoption of a scoring system at least ensures that an animal is regularly assessed, and that the possibility of pain is acknowledged. The reliability of a scoring system can be improved by tailoring the criteria used for assessment to each species and to different types of potentially painful conditions (Beynen *et al.*, 1986; LASA, 1990).

13.3. Management of Postoperative Pain

Acute pain can be alleviated by the use of centrally or peripherally acting analgesics administered systemically, by the use of local anaesthetics and by the application of supporting bandages or other means of protecting and immobilising damaged tissues. Selection of a particular regimen of treatment will vary depending upon the estimated severity and duration of the pain.

13.3.1. Opioids

If pain is judged to be moderate or severe, for example following trauma or major surgery, opioids are usually required to produce effective pain relief. A wide variety of different opioids are available, and since these have all been tested for efficacy in laboratory rats, safe dose regimens are available (Flecknell, 1984) (Table 9). What is uncertain however, is whether these dose rates, assessed using experimental analgesiometry (for example the tail flick test), are appropriate for use in postoperative pain. As our understanding of postoperative pain improves, it should be possible to revise these dose rates.

A practical problem which arises when using opioids is that the duration of action of most of these agents is under 4 hours. The more recently introduced opioid, buprenorphine, has been shown to have a long duration of action (6–12 h) in the rat (Cowan *et al.*, 1977; Dum and Herz, 1981). Clinical experience in other animal species and controlled trials in man have demonstrated that opioids are most effective in controlling postoperative pain if they are administered before pain is experienced. For this reason it is preferable to administer opioids intraoperatively, or immediately following the completion of surgery.

Table 9. Suggested dose rates for analgesics in the rat. Dose rates based on clinical experience, experimental analgesiometry (see Flecknell, 1984 and Jenkins, 1987, for review) and previously published data (Lumb and Jones, 1984; Short, 1987).

Drug	Dose
Buprenorphine	0.01–0.05 mg/kg s.c. 8–12 hourly
Butorphanol	2.0 mg/kg s.c. 4 hourly
Codeine	60 mg/kg s.c. 4 hourly
Morphine	2–5 mg/kg s.c. 2–4 hourly
Nalbuphine	1–2 mg/kg i.m. 3 hourly
Pentazocine	10 mg/kg s.c. 3–4 hourly
Pethidine (Meperidine)	10–20 mg/kg s.c., i.m. 2–3 hourly

The administration of repeated doses of opioids to maintain prolonged periods of analgesia may cause some practical problems. In man, there have been considerable advances in the technique of administration of opioids by continuous infusion and it has been repeatedly demonstrated that this provides more effective pain relief than intermittent dosing (Hull, 1985; Kay, 1987). This technique will not usually be practicable in the rat, but may be considered following major surgery, particularly if the animal has an indwelling vascular catheter. Continuous infusion of buprenorphine has been shown to provide safe and effective analgesia in the rat (Leese *et al.*, 1988).

13.3.2. Side Effects of Opioids

Some clinicians have expressed concern about the undesirable side effects of opioids. It is important to recognise that these extremely potent compounds can have undesirable properties and to use them cautiously. Concern about side effects should not, however, be used as an excuse for withholding pain relief.

Respiratory depression is perhaps the most serious consequence of overdose with opioids in man. Whilst this may occasionally be seen in animals if high doses of morphine or pethidine are administered, or if potent agonists such as fentanyl or alfentanil are used, significant respiratory depression rarely occurs following the use of mixed agonist antagonist drugs such as buprenorphine, nalbuphine and butorphanol. In circumstances when respiration is depressed as a result of thoracic surgery, alleviation of the associated pain usually results in an improvement in tidal volume and respiratory rate, and consequently in an increase in pO_2 and decrease in pCO_2.

When administered by rapid intravenous injection, some opioids, notably morphine, can cause pronounced hypotension in many species (see Short, 1987, for review). Hypotension is often a particular concern when dealing with an animal which has undergone very major surgery involving considerable tissue trauma. The more recently introduced analgesics such as nalbuphine, buprenorphine or butorphanol produce minimal effects on the cardiovascular system and so can be used safely under these circumstances (Pircio et al., 1976; Schurig et al., 1978; Trim, 1983; Schmidt et al., 1985; Cowan et al., 1977; O'Hair et al., 1988). It has been demonstrated that the partial agonist nalbuphine has a protective effect against haemorrhagic shock, and so may have a further beneficial action in addition to its analgesic effect (McKenzie et al., 1985).

A second often cited contraindication to the use of opioids is following cranial surgery. In general, opioids raise intracranial pressure and cerebral blood flow by increasing carbon dioxide tension as a consequence of their depressant effect on ventilation. As has been mentioned earlier, the degree of respiratory depression produced by opioids is rarely significant in animals and we have used opioids in our laboratory following cranial surgery in rats without experiencing any difficulties. A further problem associated with the use of opioids following cranial trauma is that they may cause sedation, pupillary dilation and bradycardia. These effects may mask signs of increasing intracranial pressure arising from postsurgical injury. If the rat is being carefully monitored, and signs of increased intracranial pressure would alter the treatment adopted, for example by prompting surgical intervention, then there may be a logical reason for withholding opioids. If no change in rat management is contemplated, then opioids should be administered, having first carried out a thorough assessment of the patient. If opioids are to be withheld, then an alternative analgesic such as a non-steroidal anti-inflammatory drug should be administered. Aspirin-type drugs, which inhibit platelet aggregation and clotting, should be avoided after surgery involving the central nervous system and paracetamol or flunixin are to be preferred.

13.3.3. Non-steroidal Anti-inflammatory Drugs (NSAIDs)

These compounds are of use in the management of mild–moderate pain in animals, and may be of particular value in circumstances when the use of opioids is contraindicated because of a particular research protocol. Virtually all NSAIDs have been evaluated in the rat during their initial development, and dose rates for these compounds are readily available (Table 10). The efficacy of these compounds on postsurgical pain has not been evaluated, however. Most of these compounds are administered by the oral route and this may cause some difficulties. Many rats find paediatric syrup preparations extremely palatable; alternatively, if fully water soluble, the NSAID can be administered continuously in the drinking water. Flunixin, which is administered by subcutaneous injection, has been shown to be a particularly effective analgesic in other species.

13.3.4. Local Anaesthetics

Extensive use is made of local anaesthetics in the management of pain in man. Although nerve blocks and even neurectomy have become

Table 10. Suggested dose rates of non-steroidal anti-inflammatory drugs in the rat. Dosage based on efficacy in inflamed paw pressure test (Vinegar *et al.*, 1976; Bartoszyk and Wild, 1989; Vinegar *et al.*, 1989).

Drug	Dose rate
Aspirin	100 mg/kg per os
Diclofenac	7.5 mg/kg per os
Flunixin	2.5–5.0 mg/kg
Ibuprofen	s.c.
Indomethacin	30 mg/kg per os
Paracetamol	2.0 mg/kg per os
	150 mg/kg per os

established techniques in clinical veterinary practice, local anaesthesia is relatively underused in laboratory animals.

Bupivacaine (Marcain) is a long-acting local anaesthetic (3–6 hours in man), with a high sensory:motor blockade ratio (Alexander and Hill, 1987). It can be used to infiltrate a wound incision to provide pain relief in the immediate postoperative period in man (Buckley,1985) and a similar technique has been used in rats (Flecknell and Liles, 1992). Specific peripheral nerve blocks can also be used to provide postoperative pain relief.

13.3.5. Recommendations

Following a minor surgical procedure such as vascular cannulation or superficial surgery involving only the skin and subcutaneous tissues, a single dose of buprenorphine (Table 9) should be administered either immediately following induction of anaesthesia, or, if a neuroleptanalgesic regimen is being used, immediately following the completion of surgery. If a shorter acting drug such as butorphanol or nalbuphine is used, the rats should be assessed 3–4 hours postoperatively to determine whether an additional dose of analgesic should be administered. If more major surgery involving penetration of a body cavity, trauma to skeletal muscles, orthopaedic procedures or cranial surgery, has been undertaken, then buprenorphine (0.05 mg/kg) should be administered every 8–12 hours for 24–36 hours. Alternatively, shorter acting opioids can be administered 3–4 hourly or by continuous infusion. In some instances, analgesics may be administered for 48 hours or longer, depending upon the severity of the surgical procedure.

An alternative to the use of opioids is to infiltrate the surgical field with bupivacaine at the time of surgery, and commence analgesic therapy 6–8 hours later. If only superficial surgery has been undertaken, then infiltration with long-acting local anaesthetic may be all that is required. Once again, careful assessment of the animal, particularly its postoperative food and water intake, should be carried out. It may not be possible to predict an appropriate analgesic regimen, but by reviewing progress of each animal in a series of experiments, the analgesic protocol used can be refined and tailored to suit a particular operative procedure.

13.3.6. Additional Aspects of Pain Management

In addition to considering whether pain may be present, it is also important to consider whether the animal is suffering "distress". Numerous definitions of stress and distress appear in the animal welfare literature. In this context the term is used to describe the state produced by adverse factors which would not usually be considered to cause pain, but are certainly unpleasant and should be avoided. For example, a cold wet environment, devoid of suitable bedding material, is likely to cause distress to rats. Similarly, states of physiological imbalance such as dehydration caused by inadequate fluid therapy would not be referred to as painful, but could cause distress.

It is also important to recognise that pain has an important emotional component, and its intensity is linked to the presence of other emotions such as fear. In man, both the intensity of pain as reported by the patient, and the requirement for analgesics to control pain, are increased by factors such as fear and apprehension (Chapman, 1985). Extrapolation of such findings to animals must obviously be carried out with great care. Nevertheless, it seems reasonable to attempt to control fear or apprehension, and to provide a comfortable environment for animals which are receiving analgesic therapy. It is therefore important to link the use of analgesics with an overall plan of patient care.

Chapter Four

SURGICAL TECHNIQUE

14. PRINCIPLES AND STANDARDS OF SURGERY

It is almost universally considered that rats rarely become infected as a result of surgery. As a consequence, surgery carried out non-aseptically but in a clean manner is routinely practised. This is completely at variance with the principles and standards of aseptic technique taught at veterinary and medical schools, where asepsis is defined as ensuring complete freedom from infection. It implies that the rat, in some way, is unique and has a resistance to infection which makes it unnecessary to apply aseptic techniques. This attitude was encapsulated rather well in a statement made in 1942 by Ingle and Griffith that "rats are less susceptible to postoperative infection than many other animals. As a matter of fact, if all aseptic precautions are disregarded but the operation is carefully performed and the animals kept in a dry cage, many of the animals will remain free from infection after most procedures." It might be expected that there is a body of scientific evidence to substantiate this apparent "fact" of resistance, but this is not so and in reality there is practically no published evidence to show, let alone suggest, that rats are uniquely resistant. So how did our current thinking about the risks of postoperative wound infection in rats arise? The following can be suggested.

First, rat surgery is carried out rarely by veterinary or medically qualified persons but in the main by biomedical scientists such as biologists, biochemists, physiologists, pharmacologists, immunologists and endocrinologists. Few will have received formal training in surgery since few such training courses are available. Surgery is learned either from colleagues, themselves not formally trained, or by self-tuition from text books, of which there are few. It is perhaps not surprising that standards of surgery do not nearly match those taught to veterinary and medical students.

Second, to biomedical scientists, the aim of any surgery in the rat is merely to prepare an animal model or tool to carry out, say, a biochemical study which is their prime interest. Therefore, an attitude is developed where there is little interest in monitoring or investigating infection. It may not even be recognised when it occurs, or the animal may simply be euthanased without a record of the infection being made.

Third, rat surgery is frequently carried out on batches of animals, sometimes numbering hundreds. Because the adoption of strict aseptic procedures under such conditions would be considered to be totally impractical and, therefore, to be avoided, any suggestion that they should be applied is ignored. It can be readily seen that as a result of such arguments, even if incidences of infection did occur, there would be no reason to publish them. Thus as a consequence of the lack of published scientific data it is difficult if not impossible to counter a statement that rats do not become infected as a result of surgery. Therefore, such a statement gets perpetuated and believed.

Is there any argument that can be made which might support a contention that high standards of asepsis are not required when undertaking surgical procedures in rats? It has been shown that the rate of infection in human operations is, in part, related to the size of the surgical wound and the duration of surgery both of which, if extensive, increase the chance of contamination by large numbers of bacteria (Blowers, 1964). Thus when comparing surgery carried out in the rat with that in larger animals such as dogs, or in humans, it is evident that the surgical incision in the rat is comparatively tiny, the duration of operations much shorter and the quantity of extravasated blood and of necrotic tissue proportionately small. However, it is well established as a general principle that for an infection to develop in a wound, a critical number of wound bacteria have to be present. In a clean wound, this is usually quoted as 10^5 bacteria per gram of tissue or millilitre of body fluid (Krizek and Robson, 1975; Swaim, 1980). This numerical concept of infection also suggests that irrespective of whether the wound is clean or "compromised" it is still the number of bacteria that is of major importance for initiation of the infection process. Since it is the absolute number of bacteria that are of importance, the number of bacteria required to establish an infection in a surgical wound in a rat will be correspondingly small.

It is also known that certain factors, such as reduced activity of host defence mechanisms which normally act to prevent the ingress of bacteria into the body, increased availability of nutrient material for bacterial multiplication (e.g. blood, necrotic tissue) and a large catchment area for micro-organisms, make the attainment of a critical number more likely (Swaim, 1980). The quantity of extravasated fluid will also be influenced by the degree of tissue trauma, and this will be related to some extent to the skill of the surgeon. Since it seems unlikely that biomedical scientists will receive the surgical training given to veterinary and human surgeons and thus will rarely achieve comparable expertise, it seems reasonable to suggest that comparatively more tissue trauma will occur during experimental surgery. However, in contrast, the duration of many operations carried out in rats is much shorter than those undertaken in man, and it is probable that the short duration of surgery leads overall to less compromise of the defence mechanisms. Thus it can be argued that some factors that are known to influence surgical wound infection are much less in evidence in rat surgery. In addition two other factors may be pertinent.

First, animal facilities, by law in many countries, have to practise high standards of hygiene. Also animals housed in such facilities start off with high standards of health—in contrast to many which are presented for treatment in veterinary practices. Thus it is to be expected and indeed has been found (Goldschmidt, 1972) that few of the common wound organisms are present in any quantity in an animal facility. Second, Donnelly and Stark (1985) have shown that rats are considerably more efficient at eliminating bacteria than other rodents, which led them to suggest that perhaps rats are better at withstanding any bacterial contamination during surgery. Such arguments support the idea that there may be a reduced likelihood of rats becoming infected during surgery in a normal animal facility.

A complicating factor which may further confuse the picture arises out of studies undertaken to evaluate the efficacy of antibiotics and antiseptics. For such work, rodent models, including the rat, are used whereby various doses of known wound organisms, such as *Staphylococcus aureus, Bacillus fragilis, Escherichia coli* and *Pseudomonas aeruginosa* are inoculated into prepared surgical wounds. The dose-related appearance of infection, usually in the form of stitch and other abscesses and of pus, is then monitored. Such studies have shown that doses as low as 10^2 for some organisms (Goldschmidt, 1972) will produce small stitch abscesses in rats indicating that rats can be infected intentionally relatively easily. This suggests that rats do not have a significant resistance to surgically acquired infection. This is supported by reports which demonstrate that the use of aseptic techniques may be essential in some operations. This is the case in cannulation of blood vessels (Popp and Brennan, 1981), thymectomy of young animals (Hard, 1975) and in situations requiring animals to be re-derived by Caesarian section to improve microbiological quality. Clinically inapparent infections have also been reported to interfere with research projects, and adoption of aseptic techniques resolved these problems (Cunliffe-Beamer, 1990).

The lack of good controlled studies on the incidence and significance of wound infections following rat surgery, as carried out in the average animal facility, makes it difficult to produce scientifically justified recommendations as to the principles of surgery that we should now be applying to rats. However, since the balance of evidence does not appear to support the contention that rats are inherently resistant to wound infections, it does not seem acceptable to adopt low standards of cleanliness and asepsis. Given the general trend in many countries to raise overall standards of animal experimentation, it seems an appropriate time to reconsider the standards applied to rodent surgery. In human operating theatres, the practice of aseptic surgery has been developed to a very advanced state, and includes not only preparation of the patient and the surgical team and their equipment, but also requires introduction of specific theatre management techniques to maximise the standards of cleanliness within the theatre area. Adoption of all of these practices in rodent surgery are impracticable and almost certainly not justified, but a number of the techniques used could easily be adapted for widespread use in many research institutes. Certain fundamental principles of aseptic surgery should be adhered to, and research workers should aim to upgrade their own technical expertise so that they can achieve these standards (see below).

When considering the approach to be adopted in a particular establishment, research workers should consider both the national and local codes of practice and legislative requirements (e.g. Animal Welfare

Act in the USA, Animals (Scientific Procedures) Act, 1986, and the relevant Codes of Practice in the UK). It is also essential that appropriate training programmes for research workers who are unfamiliar with surgical technique and aseptic procedures are made available in all institutes which undertake experimental surgery.

15. THE APPROACH TO SURGERY

15.1. Considerations before Surgery

15.1.1. Planning

When a surgical procedure with which the investigator is unfamiliar is to be carried out, careful planning is essential if the animal is not to suffer unnecessarily. It is useful to prepare a checklist of requirements, which will not only be useful to the investigator but also to any ancillary personnel who may be assisting. Such a checklist should include the following:

(1) Choice and availability of animal.
(2) Preoperative evaluation of the health of the animal.
(3) Provision of surgical and pre- and postoperative facilities.
(4) Choice and preparation of surgical instruments, apparatus and accessories.
(5) Number of assistants required.
(6) Preparation of the animal for surgery.
(7) Management of the animal postoperatively.
(8) Protocol of the surgical and anaesthetic procedures.
(9) Competence of the investigators (see below).

An important consideration is when to carry out the operation. Surgery done late in the day may require that postoperative care continues well into the evening and this could be inconvenient and lead to insufficient care being taken.

15.1.2. Qualifications and Training

Legislation does not always define the qualifications of investigators necessary for undertaking surgery, but it invariably requires that they should be competent. Those embarking on experimental surgery should have an understanding of the anatomy and known physiological peculiarities of the rat, the appropriate anaesthetic and surgical techniques available, the perioperative care required and an appreciation of the pathology which could arise from the surgical intervention. It is essential that the new investigator undergoes training, given by a person fully competent in rodent surgery and anaesthesia. Even the investigator intending to perform a very limited range of simple surgical procedures should be trained to appreciate and cope with the problems that may, and quite often do arise.

An essential part of training is practice. Live animals cannot be used in the United Kingdom for practice except in one very restricted instance, but useful information can be obtained by using a freshly killed rat. It must be remembered however that dead animals do not show the effects of bleeding, respiratory movements and tissue tension, which can complicate surgery for the novice when the first live animal is used. Practice should include handling instruments, making incisions, tying knots, closing wounds and carrying out the surgical procedure itself. The best training is achieved by discussion with an experienced colleague who has performed the operation, and by obtaining their assistance when carrying out the procedure. This, together with reading the available literature, will go a long way to preventing unnecessary manipulation of tissues which could result in unwanted haemorrhage, nerve and other tissue damage and wound infection.

15.1.3. Acclimatisation

Rats and other rodents purchased and transported from an outside source will show various physiological changes indicative of anxiety and distress, such as raised levels of blood corticosterone, raised heart rate, decreased body weight and possibly changes in immune function (Landi et al., 1982). In order not to superimpose the further stress of surgery, which could invalidate results obtained from the use of such animals in experiments, animals need to be acclimatised to their new surroundings. A period of at least 7 days is recommended. This acclimatisation period also enables a careful clinical examination to be carried out (see below) and allows preoperative assessment of body weight and food and water consumption (see 15.1.6).

15.1.4. Record Keeping

Although a clinical file is invariably kept for individual animals of larger laboratory species such as the dog, this is rarely done for the rat where information is usually kept for animals in a batch. This is a pity, as information on such things as the variability of the response of the animals to the anaesthetic used and to the effects of surgery and on the incidence of infection is not generated. Investigators should seriously consider creating individual animal files which may help in managing their results better. The type of information which could be kept is source, age, weight, sex, health status, identification, anaesthetic used and course of anaesthesia, progress of surgery, perioperative care, infection and medication.

15.1.5. Clinical Examination

Rats obtained from external sources should be healthy on arrival if they have been obtained from a reputable supplier. However, it is often necessary to check that they are and that the transport has not affected them unduly. There should be no signs of behavioural abnormality, diarrhoea or other overt signs of ill-health. Even if animals appear healthy, they may be harbouring subclinical infection. A further value of an acclimatisation period is that it may allow any subclinical infection to develop to a clinical form thus alerting the investigator before the animal is subjected to surgery. Just before surgery, animals should again be checked and allowed to go to surgery only if considered healthy.

15.1.6. Food and Water

It is generally unnecessary to withhold food and water from rats prior to surgery. However, food is often withheld overnight in operations involving the alimentary tract. Measurement of normal food and water intake is of value in assessing postoperative recovery (see 13.2).

15.1.7. *Preoperative Medication*

There is generally little need to give medicaments to rats as an aid to surgery. However, antibiotics may have a place in operations where there is a high expectation of complication by infection (e.g. in operations on the alimentary tract). Here, a long-acting broad spectrum antibiotic such as a cephalosporin (see 58) is best given some 30–60 minutes before surgery to allow peak levels to build up during the operation when there will be the greatest exposure to micro-organisms.

A single large dose of antibiotic is usually all that is required and routine administration of prolonged courses of antibiotic should be avoided to prevent the emergence of resistant organisms. Antibiotics should not be used as a substitute for good aseptic surgical technique. If the latter is employed there is rarely a need for antibiotic cover.

In some operations it may be helpful to premedicate with atropine sulphate to prevent excessive salivation if this is considered to be a problem (see Table 7).

15.2. The Operating Area

The most important aspects of the area in which rat surgery is to be carried out are that it is away from general human traffic and that it is designated only for surgery. It can vary from a specially designed and sophisticated suite of rooms to a small area on a laboratory bench. However, the latter is really only suitable for non-recovery surgery, or for simple procedures involving recovery from the anaesthetic such as biopsy of superficial tissues. Ideally, recovery surgery of whatever complexity or duration should be carried out in a separate room, kept clean with surfaces dust-free and wiped with disinfectant and with amenities such as good ventilation, scavenging for waste anaesthetic gases, good lighting and other dedicated facilities as required (Fig. 75). An anteroom for the preparation of the animal before surgery is a useful addition. The maintenance of conditions suitable for aseptic surgery is facilitated by the provision of a third room for the preparation and maintenance of instruments and sterile preparation of the surgeon and assistants for surgery.

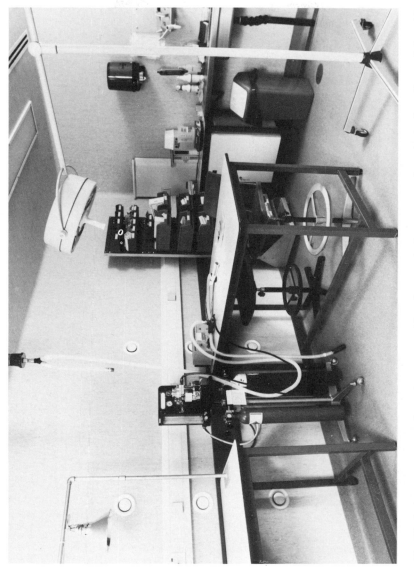

Fig. 75. A room arrangement for rodent surgery. Note the anaesthetic gas scavenging tube in the ceiling and the table with infra-red lamps to aid postoperative recovery.

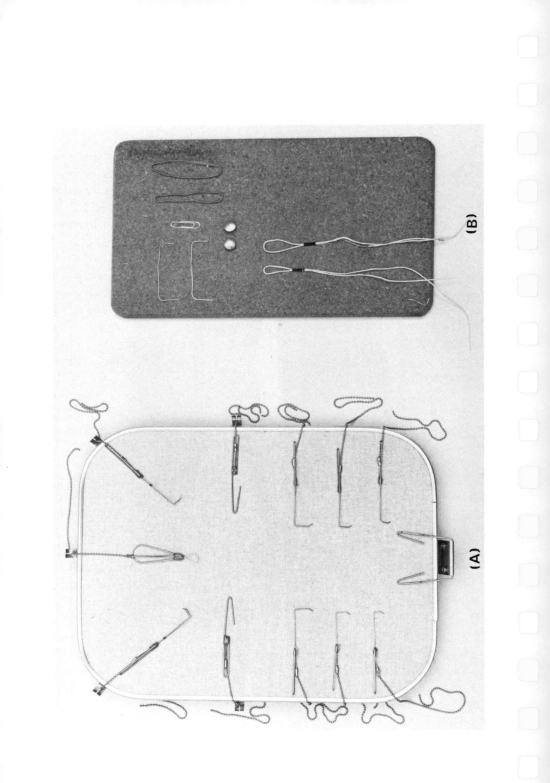

(A)

(B)

15.2.1. Operating Table and Board

Many surgical procedures require the anaesthetised rat to be held in a particular position. It may also be necessary to anchor retractors and other instruments during the operation. The small size of the rat, and the frequent lack of an assistant, makes the use of some form of operating board necessary. The simplest form of board on which to position a rat for surgery is one made out of cork of approximate dimensions 355 × 305 × 20 mm (Fig. 76). Cork is useful because of the ease with which fixing pins can be inserted. However, cork is difficult, if not impossible, to clean and disinfect thoroughly once it has become contaminated with organic matter, unless an autoclave is available. If it is not possible to autoclave the boards, then new cork boards should be used for each operating session.

A number of different techniques can be used to position the rat for surgery. Restraints to hold the limbs can be fashioned from string which is doubled through a plastic tubing slider. Alternatively, rubber bands can be used both for holding the limbs and, when placed round the incisor teeth, for steadying the head. These can then be anchored to the operating board with drawing pins (thumb tacks), dissecting pins or hypodermic needles. It is important to ensure that the ties do not occlude the blood supply to the periphery of the limb, nor should the limbs be stretched to such an extent that respiratory function is impaired. Undue tension on limbs can also cause muscle strains and nerve damage, and can produce postoperative pain and discomfort. For these reasons the minimum degree of tension should be placed on the animal's limbs. Retractors for improving visualisation of the surgical field can be made from unravelled metal paper clips. These various devices can be seen in Fig. 76. The Brookline rodent operating table with specially designed limb holders, retractors and other accessories is also shown in Fig. 76. This is no longer available commercially although it considerably aided and facilitated surgery (the authors have details of its construction which the reader can obtain). Other tables specifically for rodent use are available but these are less useful because of the lack of suitable accessories (see Equipment Index). It is important that any retractors are sterilised before use (see below).

Fig. 76. (A) The Brookline small operating board with some of the available accessories; (B) a "home-made" cork operating board with some accessories.

15.3. Equipment for Surgery

15.3.1. Instruments

These are usually chosen from those used in human paediatric and ophthalmic surgery, neurosurgery and in biological dissection and the investigators should become familiar with the more common ones (Hurov, 1978). However, lack of specific instrumentation for rat surgery requires at times the need for ingenuity in fashioning novel implements. The precise instrumental requirements for a particular operation will depend on what is available locally to the investigator, the price of the instrument and the preferences of the investigator. A basic set of instruments common to most operations would comprise of a scalpel (size 3) with a number 10 blade (Fig. 77), a pair of blunt-ended Mayo scissors, a pair of vascular scissors for blunt dissection of muscles or tissues, or a pair of small scissors for blunt dissection and a pair of pointed dressing scissors. Microsurgical scissors will be necessary if blood vessel cannulation is to be carried out (Figs 78 and 79). Two pairs of curved or straight medium fine serrated dissecting forceps and

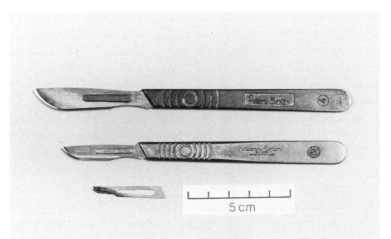

Fig. 77. Scalpel handles size 4 (top) and 3 (bottom). Size 3 is suitable for most operations in the rat. The No. 10 blade (attached to the No. 3 scalpel handle) is used for most incisions. The No. 15 blade (bottom) is used for delicate incisions.

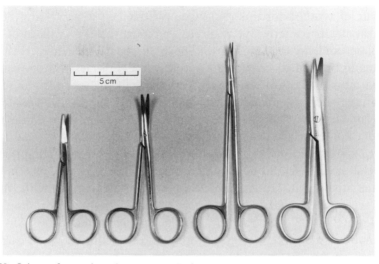

Fig. 78. Scissors for use in rodent surgery. Left to right: pointed dressing scissors for general surgical use; small scissors for blunt dissection; vascular scissors for blunt dissection—note the tapered tips which have smooth; rounded ends to avoid traumatising tissue; Mayo scissors for general surgical use and for cutting suture materials.

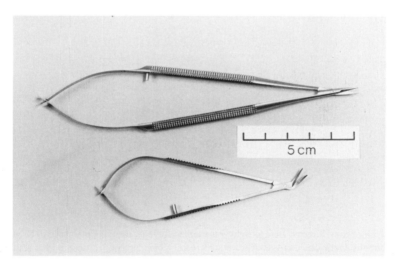

Fig. 79. Microsurgical scissors. Used for incising blood vessels during cannulation.

one pair of toothed forceps should be included in the set (Fig. 80). A more satisfactory alternative is to use DeBakey forceps in place of both the plain and toothed forceps. These forceps have fine, relatively atraumatic tips which provide a firm grip on the tissues without causing trauma (Fig. 80). Microsurgical forceps will be needed for some procedures such as cannulation of blood vessels (Fig. 81). Needle and suture materials, of which there are a variety of types (see p. 183) and a needle holder will be required for closure of the various tissue layers (Fig. 82). If the skin is to be closed with Michel clips, then an applicator for the skin clips should be included, together with additional skin clips (Michel clips). Skin staples have advantages in comparison with Michel clips (see p. 189), and if these are to be used an applicator should be included (Fig. 83). Retractors (Fig. 84), haemostats (Fig. 85) and atraumatic tissue forceps (Fig. 86) may also be required when carrying out certain operative procedures.

15.3.2. Other Apparatus and Accessories

On occasions, other apparatus may be found necessary to facilitate surgery. The need will be dictated by the type of operation to be

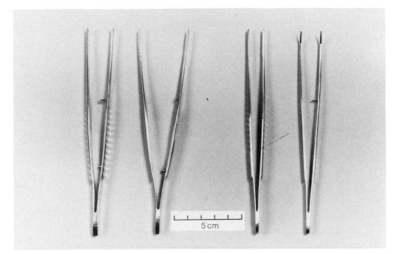

Fig. 80. Dissecting forceps. Left to right: rat-toothed forceps, these enable tissue to be grasped firmly but the teeth can damage delicate structures; plain forceps, cause minimal trauma to tissue but tend to slip; DeBakey forceps, the fine indentations on the tip enable tissues to be gripped securely with a minimum of trauma; fine-tipped forceps for delicate dissection.

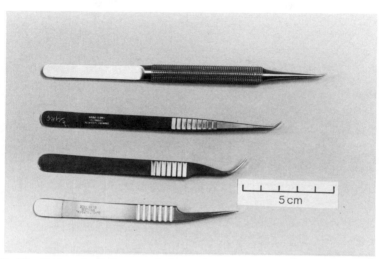

Fig. 81. Fine forceps for delicate surgical procedures. The lowest pair of forceps are vessel dilators, and have fine, rounded tips which can be inserted into blood vessels to aid passage of a vascular cannula.

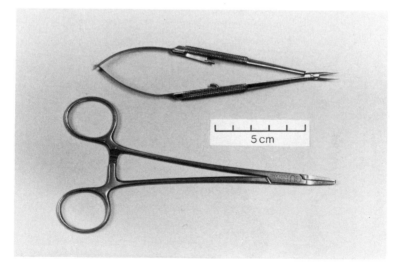

Fig. 82. Needle holders are available in a wide range of different designs. Top: microsurgical needle holders. Bottom: Ryder needle holder, a useful general purpose design suitable for most procedures in the rat.

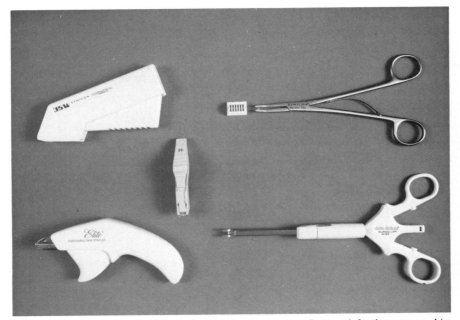

Fig. 83. Stapling apparatus. Top left: Ethicon skin stapler. Bottom left: Autosuture skin stapler. Centre: 3 M skin stapler. Top right: Ethicon ligaclip ligating stapler. Bottom right: Autosuture ligating stapler.

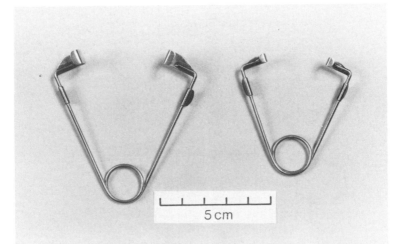

Fig. 84. A variety of different types of self-retaining retractors are available. The two shown are human paediatric instruments (Aberdeen's spring retractors).

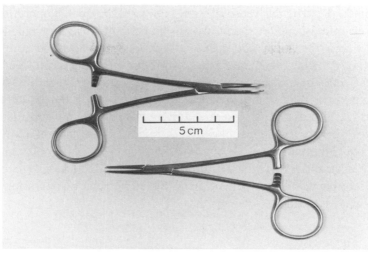

Fig. 85. Haemostats are required to clamp vessels prior to ligation.

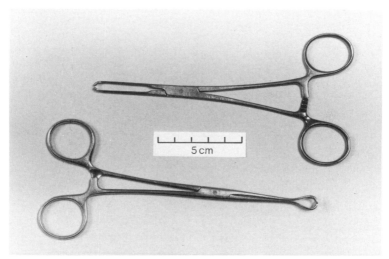

Fig. 86. Atraumatic forceps to clamp and retract tissue. Top: Allis tissue forceps. Bottom: Babcock's tissue forceps.

performed. A particularly useful item is a zoom binocular operating or dissecting microscope. A model which is of the greatest convenience is one in which the zoom and focus adjustments are operated remotely with a foot control, relieving the need to remove the hands and perhaps the eyes from the surgical site. These microscopes are generally much more expensive than manually operated ones but well worth the extra expense. A microscope is essential for microsurgery and considerably facilitates surgery involving close work such as cannulation of blood vessels and lymphatics.

Many accessories may be required in surgery, such as gauze pads, drapes, cotton wool, etc. (Figs 87 and 88). The type of operation will dictate what is needed.

15.3.3. Lighting

A good bright source of light is an absolute necessity. Inexpensive lighting can be obtained by the use of an ordinary light bulb held in an angle-poise lamp which can be variously positioned. Considerably superior lighting at a modest cost is obtained by using a fibre-optic flexible cold illuminating system (see Equipment Index). With this, precisely positioned spot and diffuse lighting can be obtained with the

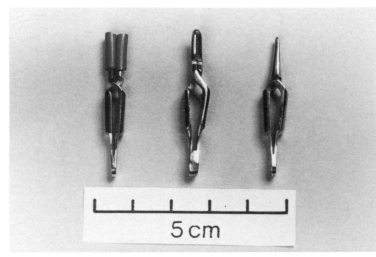

Fig. 87. Bulldog clamps for occluding blood vessels during cannulation (other, finer clamps, e.g. Scoville–Lewis, are also available, see Fig. 126).

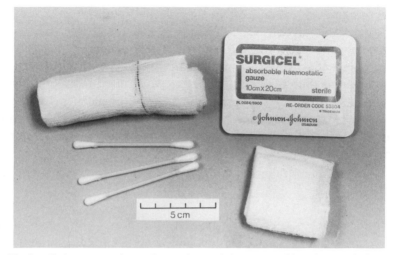

Fig. 88. A rolled gauze swab can be used as a bolster to position the rat during certain surgical procedures. Cotton tipped applicators can be used to remove small quantities of blood from the surgical field, and also may be used for blunt dissection of some structures. Surgical haemostatic gauze may be required to control haemorrhage, e.g. after partial nephrectomy. Small gauze swabs are most suitable for general use in the rat.

advantage that the light source does not produce any significant heat. The best lighting system, though the most expensive, is that specifically designed for human or veterinary operating theatres. There are various types with either a single or multihead light source and a suitable one, either floor, wall or ceiling mounted, can easily be provided for rodent surgery (Fig. 75).

15.4. Preparation and Sterilisation of Instruments, Apparatus and Accessories

The operating area should be clean, and all flat surfaces and surfaces of large apparatus should be wiped over with disinfectant. A large sterile cloth or paper towel (drape) should be placed on the operating table which should be of the free-standing type to allow access from all sides. The operating board should be sterile or have been cleaned with disinfectant. Surgical instruments, small apparatus and accessories must be sterile and either arranged on the sterile table drape around the operating board as convenient, or in a sterile instrument tray close at hand. A possible layout can be seen in Fig. 89.

The removal of dirt and micro-organisms from items to be used in surgery is essential. Initially, instruments should be scrubbed clean in water and dried. Subsequently, if they are to be used in aseptic surgery they will need to be sterilised for the removal of all forms of micro-organisms. This can be achieved in a number of ways depending on the extent of cleanliness required and on the type of instrument, apparatus or accessory to be cleaned.

15.4.1. Heat

The preferred method of sterilising instruments is to use steam under pressure in an autoclave (a pressure cooker can be used if an autoclave is unavailable). A pressure of 15 lb/in^2 for 20–30 minutes usually suffices unless a vacuum autoclave is used when shorter times can be employed. Apparatus not affected by steam and high temperatures, such as glass beakers and glass syringes, can also be sterilised in this way. These are

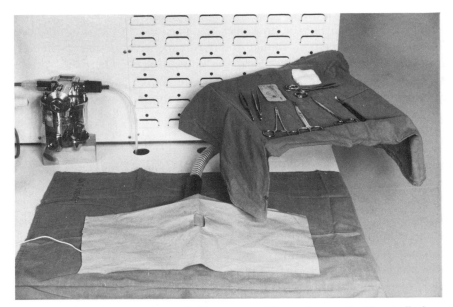

Fig. 89. Small overhang tray for surgical instruments. The tray is covered with a sterile drape before use.

all specially packaged in commercially available packaging material before sterilisation and for maintenance subsequently in a sterile state (see Lane, 1989). Items which must not be exposed to moisture, e.g. paraffin-gauze and paper towels, as well as instruments and glass and metal apparatus can be wrapped in metal foil and sterilised by dry heat in an oven. The temperature used is approximately 160°C for 1 hour. The tips of instruments can be sterilised by placing them in a hot bead steriliser which exposes the instruments safely to 250°C (see Equipment Index).

Where the removal of the vegetative forms of micro-organisms but not necessarily all spores is acceptable, this can be achieved by boiling. Clean instruments and other small items including plastic items (e.g. catheters) can be boiled in water containing 2% sodium carbonate (washing soda) for 10 minutes. Special apparatus to facilitate this is available and it is a useful and safe method that could be adopted by any laboratory. It has the major disadvantage that swabs and drapes cannot be prepared in this way since they will be unsuitable for use if soaked in water. If other methods of sterilisation are unavailable, then swabs and drapes can be purchased in presterilised packs.

15.4.2. Irradiation

Other methods of sterilisation include gamma irradiation which is most often used by commercial companies for instruments and apparatus made of plastic (e.g. syringes, catheters) but is not limited to these. The sterilised items are provided prepackaged in an outer protective wrapping.

15.4.3. Ethylene Oxide

Sterilisation by exposure to ethylene oxide gas can also be used, and small sterilisers are available commercially (see Equipment Index). This method is particularly suitable for sterilisation of catheters and cannulae. Ethylene oxide is explosive and inflammable, and toxic, so that special care must be taken in its use. Many commercially produced sterilisers are equipped with an extract system to enable controlled venting of any remaining ethylene oxide to prevent contamination of the room. Most systems require the material to be sterilised to be exposed for about 12 hours, and then allowed to stand in air for a further 12 hours to allow the gas to dissipate from the packing materials.

15.4.5. Chemical Disinfectants

As an alternative to the methods described above, instruments can be completely immersed in disinfectants for a minimum period dictated by the disinfectant used. Common examples of such solutions are 70% ethyl alcohol, industrial methylated spirits 74% overproof, 0.1% benzalkonium chloride in water or 70% alcohol, 0.5% chlorhexidine glucorate in water or 70% alcohol, or 2% cetrimide in water or 70% alcohol. Items should be left in these disinfectants for at least 15 minutes initially if no other instructions are available. A number of products are marketed specifically for the purpose of cold sterilisation of instruments (e.g. Asep, see Equipment Index). Before using instruments or plastic items which have been immersed in disinfectant, they should be rinsed in sterile water or saline to prevent the carry-over of potentially irritating chemicals into the wound. Care must be taken not to leave certain instruments immersed in fluids such as Asep for too long. In particular, instruments with tungsten inserts are easily damaged by this solution.

Apparatus which cannot be immersed in disinfectant should be wiped over with disinfectant to achieve a reasonable degree of cleanliness. Accessories which cannot be cleaned with disinfectants, such as gauze pads, cotton wool and surgical drapes, should ideally be prepared using a vacuum autoclave, or purchased prewrapped in sterile packs.

15.5. Considerations during Surgery

The main aims of surgery are, first to carry out the surgical procedure skilfully with the minimum of risk and disturbance to the patient, second to carry out surgery without the accompaniment of infection.

15.5.1. Aseptic Technique
15.5.1.1. Preparation of the surgeon. When working in a dedicated experimental theatre suite, the investigator may be required to remove his or her outer clothing and wear a lightweight "scrub suit". If the surgery is to be carried out in a room designated only for rodent surgery,

then a number of procedures may be adopted to minimise the risk of contamination of the operative field. The surgeon should wear a clean surgical cap and face mask. He should next remove any jewellery and watches and wash hands and arms thoroughly with antiseptic soap (e.g. Hibiscrub or Betadine surgical scrub) and a scrubbing brush. Using a commercially available scrub brush which is presoaped (E-Z Scrub, see Equipment Index) is a convenient way of doing this. The hands and arms should be dried with a sterile towel and the investigator should put on a sterile gown and gloves (see Lane, 1989).

15.5.1.2. Preparation of the animal. Whilst the investigator is getting prepared, an assistant should anaesthetise the animal and remove the hair from the surgical site using electric clippers (see Equipment Index). It is rarely necessary to proceed further but if necessary, fine hair can be removed using a razor, taking great care not to abrade or cut the skin, or a depilatory (available from pharmacists) can be used. Removal of hair from too large an area of the body should be avoided to prevent unnecessary heat loss (about 2 cm on either side of an abdominal incision on an adult rat is sufficient). This should be carried out on a table separate from that on which the operation is to be performed (an adjacent room is preferable). The hair which has been removed can be disposed of conveniently by using a small vacuum cleaner, of the type sold for cleaning the interiors of automobiles. The animal should then be placed on the operating board.

15.5.1.3. Positioning the animal. Most operations in the rat are performed with the animal on its back (dorsal recumbency). In this position the limbs may need to be stretched out using slings or ties which will need to be fixed to the operating board. The limbs, particularly the front ones, must not be overstretched otherwise respiration may be compromised. For most surgical procedures, the investigator will find it more convenient to sit at the table rather than stand. This will allow the forearms to rest comfortably on the operating table, and so reduce tremor and stabilise the hands and wrists.

15.5.1.4. Use of antiseptics. The area of the incision site should either be sprayed with antiseptic using an aerosol or atomizer and the excess removed immediately with gauze, or alternatively a gauze, or cotton wool pad soaked in antiseptic should be wiped over the area. Swab the surgical site working from the middle outwards. Although 70% ethyl or isopropyl alcohol have been traditionally used for many years, better antiseptic cover for longer periods is usually obtained by the use of relatively newer agents. The following commercially available antiseptics will be found suitable: chlorhexidine glucorate 0.5% (w/v) in 70% isopropyl alcohol (e.g. Hibitane), benzalkonium chloride 0.1% (w/v) in 70% isopropyl alcohol (e.g. Roccal) and a 10% (w/v) alcoholic or aqueous solution of Povidone-iodine (e.g. Betadine alcoholic or antiseptic solution or spray). The latter has gained much popularity in recent years. It is effective against a wide range of vegetative organisms and although containing iodine is non-staining. The antiseptic should be allowed to remain on the skin for at least 3 minutes to act on surface bacteria before the incision is made.

15.5.1.5. Draping the animal. Surgical drapes are used to avoid instruments or exposed viscera coming into direct contact with parts of the animal which have not been shaved and prepared for surgery. A sterile cloth, paper or adhesive plastic drape (e.g. Steridrape) should be placed over the animal leaving only the head (unless it is the site of surgery) and the surgical site exposed. For rats, the most useful type of cloth or paper drape is one in which a hole has been cut corresponding to the area of the surgical site (see Fig. 90). If an adhesive plastic drape has been used, the skin incision can be made through it. These drapes do

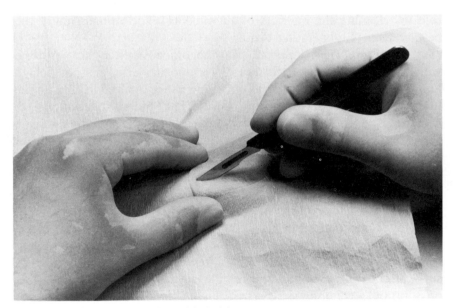

Fig. 90. Use of a scalpel to make a skin incision.

not adhere well to wet surfaces and some of the antiseptic may need to be removed to get good adhesion.

16. SURGICAL TECHNIQUE

16.1. Surgical Approach and the Use of Instruments

16.1.1. The Incision

Once the skin has been prepared with antiseptic, the skin incision can be made with a scalpel. The scalpel should be held at an angle of about 30° to the skin. The skin on either side of the chosen incision site should be placed under gentle tension using the thumb and forefinger of the operator's other hand (Fig. 90). The scalpel should be drawn smoothly along the line of the incision. Avoid taking several hacks or stabs at the skin as this will produce an uneven, irregular wound that will be difficult to suture and will heal slowly. The depth of the cut should be such that only the skin is incised. The pressure that needs to be applied to the scalpel to achieve this can be judged by observing the depth of the incision behind the scalpel as it is drawn across the skin. Practice and experience are needed to produce a clean, straight incision of an even depth. After incising the skin, the scalpel should be placed on one side. In virtually every operation, further cutting of tissues is best carried out using scissors. Before proceeding further, some anatomical knowledge of the surgical field is essential, and this should have been gained by careful dissection of post-mortem specimens prior to undertaking surgery in a living animal. Using such anatomical knowledge, incisions can be made that avoid major blood vessels and nerves, e.g through the linea alba along the midline of the abdomen to gain access to the abdominal cavity. Use of such techniques minimises haemorrhage and tissue damage, and so is an important aspect of surgical training.

16.1.2. Tissue Handling

Once surgical exposure of the anatomical structure of interest has been achieved, subsequent procedures will depend on what the investigator is trying to achieve. However, irrespective of the type of operation, the outcome of the surgery and the welfare of the animal are closely dependent on adoption of good surgical technique. Tissue trauma can be minimised by gentle unhurried handling of tissues. This is a major requirement of good surgery and it cannot be too heavily emphasised that expertise is only gained by practice. Excessive use of retraction or inexpert handling of instruments may damage tissue which then becomes necrotic and provides excellent nutritive material for bacterial growth. Only tissues that can be seen clearly should be cut, having first examined the proposed incision site carefully to ensure that no blood vessels or other vital structures are likely to be damaged. Scissors can be used both to cut tissues and for blunt dissection. There is no need to place tissue under tension in order to cut it with scissors. For maximum control, the thumb and ring finger should be placed through the handles of the scissors, and the index finger positioned near the hinge joint (Fig. 91). Left-handed use of scissors reduces the shear and torque produced by the blades. The effectiveness of cutting can be improved by pulling with the thumb and pushing with the ring finger. This seemingly awkward technique may already have become second nature to left-handed persons (one of the authors is left-handed). It is possible to purchase left-handed scissors if problems are experienced.

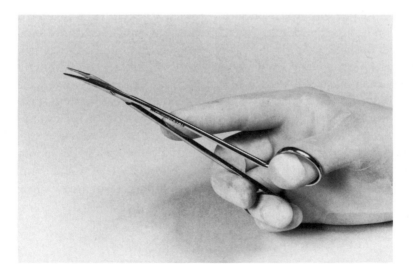

Fig. 91. Hold scissors with the index finger positioned close to the hinge joint to stabilise the instruments during use.

Scissors are available with either straight or curved blades and with blunt or sharp tips (Figs 78 and 79). Sharp pointed scissors can be used for piercing tissues, enabling a cut to be made. They have the disadvantage of causing potentially disastrous injury if inadvertently stabbed into a major vessel or hollow viscus during surgery. To avoid this problem, use blunt ended scissors whenever possible and use pointed scissors only for making an initial incision, for example when entering the abdomen.

Blunt dissection is to be preferred to cutting of tissue wherever possible. To blunt dissect between tissue planes, blunt ended scissors should be inserted closed, then opened gently to separate the tissues (Figs 92 and 93). When using forceps to hold or retract tissues during surgery, one blade should act as an extension of the thumb, the other as an extension of the opposing fingers (Fig. 94). As mentioned earlier, DeBakey forceps are recommended for routine use.

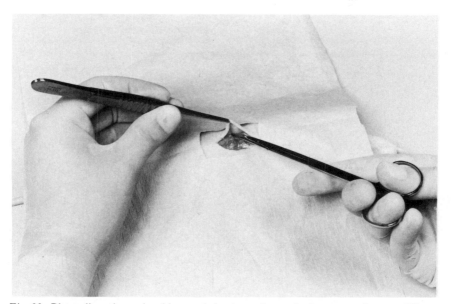

Fig. 92. Blunt dissection using blunt-ended scissors is required as part of many different surgical procedures. The scissors should be inserted, with their blades closed, between the tissues to be dissected.

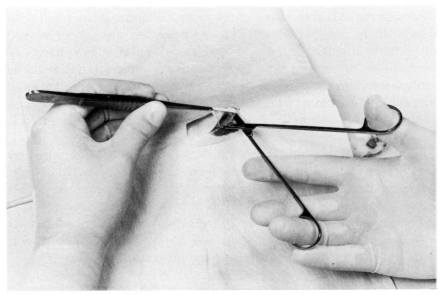

Fig. 93. Blunt dissection: after insertion of the scissors, the blades are gently opened to separate the tissues using the natural tissue planes.

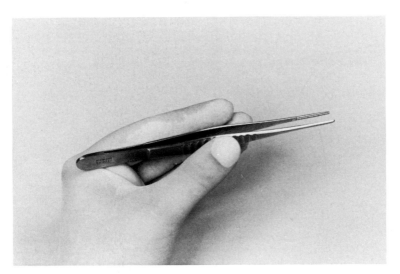

Fig. 94. Hold forceps so that the blades act as an extension of the thumb and the opposing fingers.

16.1.3. Haemostasis

Some bleeding during surgery is inevitable but blood loss must be minimised at all times. The ability to ligate blood vessels or to use a cautery if required or to use other established methods to stop bleeding must be practised. Too much blood loss will obscure the surgical area making it difficult to continue or at worst will cause hypovolaemia and cardiovascular failure ("shock"). If haemorrhage is occurring from an identifiable vessel, then this should be ligated, and this is most easily achieved by first clamping the damaged section with curved haemostats (Fig. 85). After grasping the vessel, a ligature can be passed around the forceps and tied. The forceps can then be removed. If complete ligation of the vessel is not possible, for example if a small hole has been made in a major artery or vein, then pressure should be applied using a small swab or piece of haemostatic material (see Equipment Index). After 2 or 3 minutes, the pressure can be removed and the haemorrhage will probably have ceased. Extravasated blood and blood clots must be removed as these also provide an ideal growth medium for bacteria and so predispose to wound infection. When removing blood or blood clots, or trying to control capillary haemorrhage (the slow "ooze" seen from cut tissues), apply a swab firmly, then lift it off. Do not try to clear a surgical field by using the swab with a scrubbing action, since this will dislodge any small clots which have formed and will start further haemorrhage.

16.1.4. Wound Irrigation

Some operations, notably those in which the integrity of the lower alimentary tract is breached, are inherently "dirty" and spillage of the contents into the abdominal cavity may produce serious or fatal infection. Any such spillage, or other contamination (e.g. excessive blood), should be removed by copious washing using warm (37°C) sterile saline (0.9%). It is preferable carefully to isolate the area of bowel which is to be incised by the use of swabs soaked in warm sterile saline. This will reduce any contamination of the peritoneal cavity.

16.1.5. Other Considerations during Surgery

Every attempt must be made to minimise contamination of the operative field during surgery. Inexperienced investigators should take care to avoid simple actions such as wiping perspiration off the brow with one's sterile gloved hand! This is a task for the assistant. If gloves or other sterile attire become contaminated, they will have to be changed for new sterile ones. If an instrument becomes contaminated and a sterile replacement is not at hand then it should be disinfected or heat sterilised using one of the methods described earlier.

At the conclusion of surgery, removal of all materials from the wound such as gauze or cotton wool swabs must be ensured. The wound can then be closed using a sterile needle and suture material.

16.2. Wound Closure and Suturing

The joining together of cut edges or the closure of a surgical incision can be carried out using a variety of different suture methods. The actual method used will depend on the type and location of the tissue to be sutured as well as the preference of the investigator. Suturing technique is described in all general books on surgery (e.g. Lang, 1982; Anderson and Romf, 1980). The components of a suture method are the needle, the suture material, the knot and the type of stitch.

16.2.1. Needles

These are available in a variety of thicknesses, sizes and shapes. They are described in terms of their size or gauge, the shape of their body and the shape of their cross section. The most commonly encountered shapes are straight, half curved and curved. Straight or half-curved needles are used for suturing tissues which are readily accessible, such as skin. Curved needles are used for suturing deeper, less accessible tissues. The body of the needle will usually be either round in cross section or triangular. Round-bodied needles are used for suturing soft tissues such as muscle, bowel and fat. Needles with a triangular cross section are referred to as "cutting edge needles" and are used to suture tougher tissues such as skin.

Needles either have the thread already bonded to the eye ("swaged" or atraumatic needles) or need to be threaded before use. Needles with the thread already bonded in place are highly recommended. They do not become unthreaded at critical moments and because they are disposed of after use, the needle used will always be sharp. Since the thread is bonded cleanly to the body of the needle, the hole made when it is passed through the tissues does not exceed the diameter of the body of the needle, unlike the effect produced when a knotted suture is dragged through tissue.

In rat surgery it is rarely necessary to use a wide range of different needles and a half-curved or curved round-bodied intestinal needle about 20 mm long should prove to be the most convenient for suturing soft tissue. A similar smaller (10 mm) needle is preferable when suturing bowel. A half-curved or straight cutting edge needle (20–30 mm) should be used to suture skin.

It is preferable to hold the needle with a needle holder, and some practice will be required to master the use of this instrument. Use of a needle holder allows considerably greater precision in positioning sutures and ligatures, and aids suturing of less accessible structures. When using a needle holder, the thumb and ring finger are placed through the instrument, and the index finger is placed close to the hinge joint to provide stability (Fig. 95). This type of needle holder (e.g. Ryder) will be found easier to master than the types which incorporate locking ratchets or a scissor blade for cutting sutures. The needle should be held between the halfway point and the end nearest to the suture attachment. This allows more of the needle to protrude from the tissues after it has been inserted, and so it will be easier to grasp and pull clear. When suturing tougher tissues such as skin, it may be necessary to grasp the needle near to its tip. This should rarely be necessary when operating on a rat unless the needle has become very blunt.

16.2.2. Suture Materials

Suture thread is made from a variety of materials and is classified as either absorbable material, which is absorbed in the body within 60 days, or non-absorbable, which eventually becomes encapsulated in fibrous tissue. Absorbable sutures loose their tensile strength before being absorbed. Sutures are also described as either single-stranded

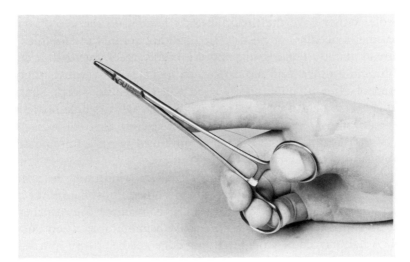

Fig. 95. Hold a needle holder with the index finger positioned close to the hinge joint to stabilise the instrument during use and so enable accurate placement of sutures.

("monofilament") or multi-stranded ("braided"). The most widely used absorbable suture material is catgut, which is prepared from sheep submucosal tissue. Plain catgut is absorbed from the suture site in about a week, chromic catgut in 2–3 weeks. Catgut is used for suturing soft tissues and generally speaking chromic catgut is preferred because it retains its tensile strength for longer. Alternatives include polyglycolic acid ("Dexon"), a multifilament suture which is stronger than catgut, and easier to knot; and polyglactan ("Vicryl"), another multifilament suture which is also easy to knot and which is even more resistant to hydrolysis than polyglycolic acid sutures. Both of these materials produce a smaller inflammatory response than catgut, and are considerably easier to use, but are more expensive. Both materials can be used to suture both internal structures and skin.

Non-absorbable materials that are in widespread use include braided silk and a variety of monofilament sutures made from plastics. The most commonly used monofilament sutures are nylon ("polyamide") and propylene polymer ("prolene"). All of the plastic-based materials are relatively difficult to handle and knot successfully. The easiest to use is the polyester "Mersilene". Braided silk is slightly more expensive than the monofilament sutures, but it is easy to handle and is useful for suturing skin. For microsurgical procedures, monofilament nylon is used since it is possible to produce very fine sutures of an even diameter.

Suture materials are provided commercially in individual sterile packs which have an outer wrapper and an inner packaging. The suture is commonly supplied with the needle attached. Non-sterile packs are also available, as are large cassettes or spools of suture material. Catgut or synthetic absorbable materials cannot be resterilised by autoclaving, but silk and synthetic non-absorbable sutures may be resterilised in this way.

Suture materials are available in various thicknesses. A progressive decrease in thickness is indicated by size 0, 2/0, 3/0, etc. to size 11/0. Increasing thickness is marked by sizes 1, 2, 3 and greater. For the majority of suturing and ligating purposes in the rat sizes 3/0 or 4/0 are the most generally useful.

16.2.3. Making Knots

The basic knot in rat surgery is the reef knot. To practise forming this knot, the two ends of a piece of thread should be held with both hands. The left hand is passed over the right hand and the end of the string in the left hand is passed through the loop so formed. Both ends are pulled tight to form the first half-hitch. The process is repeated but now passing the right hand over the left hand and the right hand string through the loop. Pulling both ends forms the second half-hitch and completes the reef knot. Forming a third half-hitch will produce a double reef knot (sometimes called a triple throw knot) which provides added strength and safety. It takes only a little imagination to transfer the description above to a piece of thread attached to a needle which has been passed through the two cut edges of an incision (Fig. 96). To close any incision with thread it is necessary to start the suture with a reef knot. Since the investigator will be working with forceps in most cases it would be useful to form a reef knot using forceps and the method is shown in Fig. 97.

If the incision is closed with a continuous stitch, it will be necessary to end the stitch with a knot. If a needle with an eye has been threaded only once then a reef knot can be formed directly as in Fig. 98 using fingers or forceps to tie the knot. This is possible because the thread is able to slide in the eye of the needle. If the needle has been threaded

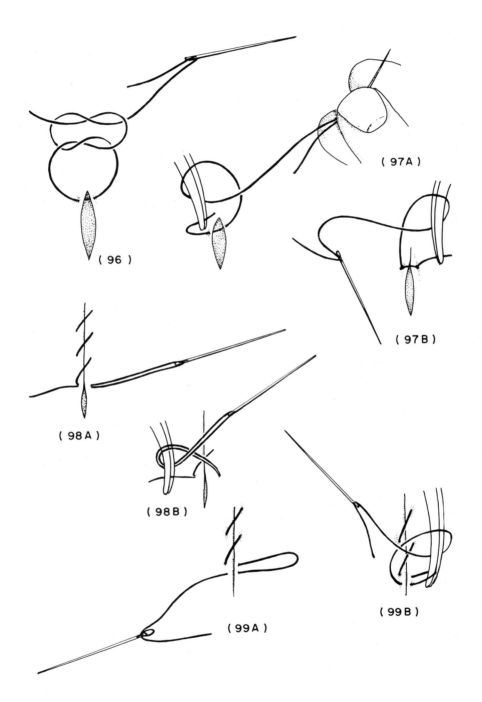

(96)

(97A)

(97B)

(98A)

(98B)

(99A)

(99B)

twice or the suture material has been bonded to the needle, the thread cannot slip and a knot is made after first passing the needle back through the two cut edges of the incision leaving a looped tail of thread, which is then treated as if it were a single piece of thread. Using fingers or forceps a reef knot is then formed (Fig. 99). At least 5 mm of thread should be left attached to the knot after cutting off the excess, to prevent the knot from unravelling.

The *ligation* of blood and other hollow vessels is also made with thread employing a reef knot. However, recently ligation to occlude hollow vessels has become possible using metal staples which are crimped against the vessel by specially designed applicators (Fig. 83). This procedure is considerably quicker than ligation using thread and can be recommended.

16.2.4. Sutures (Stitches)

There are a variety of suture patterns designed to cope with the different aspects of surgery and the investigator may want to consult a good human or veterinary textbook (e.g. Lang, 1982; Knecht et al., 1987). Which one to use is often a matter of preference for the investigator, but for the surgical procedures described in this book the choice can be narrowed down to three: (i) the continuous stitch, (ii) the simple interrupted stitch, and (iii) the horizontal mattress stitch. These are shown in Fig. 100. A rule in surgery is that a stitch which everts two cut edges should be applied to vascular anastomoses where it is important that the endothelial surfaces are opposed to prevent clot formation, and also to suturing of the skin. In other cases the stitch should bring the edges together in their natural position. However, for most general closure procedures of soft tissue in the rat the continuous stitch is the one of choice. It is simple and quick to carry out and allows wound

---

Fig. 96. A reef knot (in perspective). *Fig. 97.* (A) Forming the first half-hitch of a reef knot using forceps; for the tying of ligatures, particularly in a restricted space, the fingers are replaced by a second pair of forceps; (B) forming the second half-hitch with forceps. *Fig. 98.* Forming a reef knot to close an incision when the needle has been threaded only once, note that (B) shows only the first half-hitch. To complete the knot the second half-hitch has to be formed as in Fig. 97B. *Fig. 99.* Forming a reef knot to close an incision when the needle has been double threaded or bonded to the thread. A loop is formed and is used to make the first half-hitch, the second half-hitch is made as in Fig. 98A.

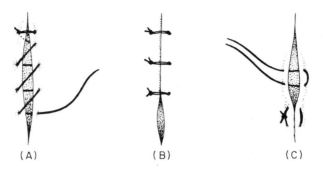

Fig. 100. (A) The continuous stitch, (B) the simple interrupted stitch, (C) the horizontal mattress stitch.

healing to proceed normally. It is often said as a note of caution or to contraindicate its use that if it is cut or broken the continuous stitch will completely unravel allowing the whole wound to open. In reality this rarely, if ever, happens providing the correct thickness of suture material has been used and the suture has been inserted sufficiently far below the cut edge so that it will not tear through it. In all cases, the needle should be inserted about 5–8 mm below the edge of the incision during suturing. The interval between the point of entry of the needle each time should be about 8–10 mm. The thread should not be pulled so taut during suturing that the tissue is strangled and damage occurs. The two edges of the incision should be gently but firmly opposed. Some investigators prefer to use the interrupted stitch or the horizontal mattress suture which everts the cut edges, as they believe that the interrupted sutures provide increased strength. This must be balanced against the considerably prolonged time for wound closure. However, for more specialised aspects of surgery, e.g. microvascular surgery, the use of interrupted sutures may be mandatory.

In large laboratory animals closure of wounds, for example an abdominal incision into the peritoneal cavity, takes place in four or more layers. Often the peritoneum is stitched, followed by the abdominal muscle, followed by apposition of subcutaneous tissue and fat ending finally with the skin. In the rat all these separate tissues are thin and closure of a midline abdominal incision is a two-step procedure, involving suturing of the peritoneum and muscle together, followed separately by closure of the skin. However, in obese animals it is also advisable to appose the subcutaneous fat with a continuous stitch. This minimises the risk of fluid accumulation postoperatively, which would form a focus for infection.

The skin may be closed either by using interrupted mattress stitches, using either 3/0 or 4/0 monofilament nylon or braided silk, or wound clips, or tissue adhesive. In some cases it may then be useful to protect the wound after closure using an aerosol plastic spray covering (e.g. Op-site—see Equipment Index).

Opinion varies as to the advantages of the different methods of closure of the skin. Suturing can be carried out quickly and easily once the technique has been mastered, and the majority of rats tolerate skin sutures well. Some animals rapidly remove their skin sutures, and this may occur because of overtightening of the sutures which causes pain and irritation at the wound site. It may also occur as a consequence of normal grooming behaviour. As an alternative to suturing, the wound may be closed using Michel clips (Fig. 101) or skin staples (Fig. 102). Michel clips are applied using special applicator forceps and should be placed about 8–10 mm apart. Michel clips are supplied in packs and have to be sterilised before use. A major disadvantage of these clips is their tendency to be overcrimped, especially when used by the inexperienced investigator, leading to local tissue damage and necrosis, providing a possible focus for bacterial growth. Skin staples are applied using an applicator gun and cannot be overcrimped because of the method of application. They also allow the normal postoperative tissue oedema, which is one of the initial stages of wound healing, to occur unhindered and this is believed to be beneficial. The use of staples in rodents has been shown in studies on intentional infection to produce considerably less infection at the point of application than silk or synthetic suture materials (Stillman *et al.*, 1984) which further indicates their suitability for use. Thus there would seem to be clear advantages in the use of staples to close skin incisions, though they are more expensive than Michel clips. When applying skin clips or staples, wound edges should be opposed neatly, taking care not to trap hair between them (Fig. 102). The edges should be pulled up well away from the underlying musculature. Very occasionally rats may remove one or more clips or staples with their teeth and these should be replaced as soon as possible. Clips and staples should be removed about 8 days after application using the special tool that is available. Anaesthesia is not required to do this. Although there seems to be little harm in leaving the clips in longer, this does have the potential for increasing the

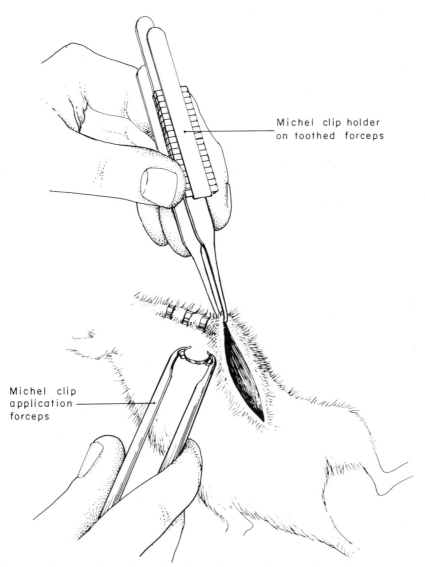

Michel clip holder
on toothed forceps

Michel clip
application
forceps

Fig. 101. Closing a skin incision with Michel clips, note that the skin is pulled upwards away from the underlying abdominal muscles during application of the clips.

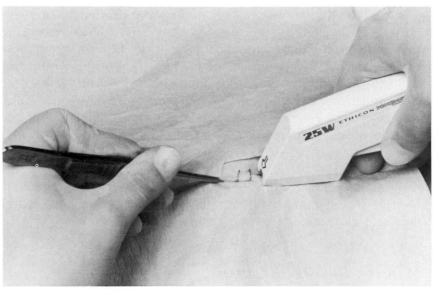

Fig. 102. Application of skin staples.

possibility of local infection. A number of different manufacturers produce staples, and the design varies considerably (Fig. 83, and Equipment Index). Because of the difference in skin thickness between humans and rats, some designs of staples are unsuitable for use in rodents.

Tissue adhesives (see Equipment Index) have been reported as being suitable for skin closure in rodents, and they have the advantage of being rapidly and easily applied. In the authors' experience, wound opening following excessive grooming and cleaning by the animal occurs more frequently when tissue adhesives are used in comparison with the other techniques described above. They are, however, useful for the emergency repair of a skin wound if anaesthesia to allow resuturing is contraindicated.

16.3. Batch Surgery

If more than one animal is to be operated on, the same set of instruments may be used providing major contamination during the first operation has been avoided, otherwise a fresh sterile set of instruments must be prepared. It is advisable to adopt procedures, such as placing all instruments back onto a sterile instrument tray when not in use, and avoiding touching the outside of the rat with the tips of instruments, to help to minimise contamination. Similarly, instruments such as the scalpel which are used to cut skin should be kept separate from instruments used to handle deeper tissues. After completion of the procedure on the first animal, the instruments can be wiped clean of organic matter and placed in a cold sterilising solution such as Asep (Equipment Index—see also 15.4.5). They should be rinsed in sterile water or saline before reuse. A convenient way of managing the situation is to have two sets of instruments, one set of which is maintained in disinfectant while the other set is being used.

16.4. Acute, Non-recovery Surgery

If an experimental animal is not to be allowed to recover from the anaesthetic, then the development of infection is not generally an issue. Consequently, techniques designed at reducing the risk of wound infections, as described above, are not necessary. However, the techniques used should incorporate basic hygiene measures and at the very least instruments should be clean. The incision site should be shaved and wiped with antiseptic simply as an aid to making a controlled and accurate incision. Good surgical technique is still required so as not to damage tissue or cause unnecessary bleeding which could have a detrimental effect on the animal even in the short term.

17. POSTOPERATIVE RESTRAINT

In certain cases where, for example, blood vessels or other ducts have been intubated for the purposes of fluid collection or for the infusion of fluids over a long period, it will be necessary to prevent the rat from biting through or pulling out these tubes. Several ways to restrain the animal from doing this are available and some of these will now be described.

17.1. The Elizabethan Collar

The simplest way to prevent an animal from biting a cannula or other piece of apparatus placed about its body is to make it wear an Elizabethan collar (Fig. 8). These are commercially available (see Equipment Index). The method is not suited to all procedures, however, and other, more restrictive techniques may sometimes be necessary.

17.2. The Bollman Cage

Another simple method of restraint is to place the rat in a Bollman cage, a design of which, similar to that of the original, is shown in Fig. 103A. Rats can be kept in a Bollman cage for several days. It does not take much imagination to realise that such animals become highly stressed; it is known that if no movement is allowed at all, rats in such cages acquire gastric ulcers within 24 hours. Because of the stressful effects of this type of restraint, this approach should be considered the least desirable means of protecting catheters, and it has been discarded by many laboratories in favour of more humane techniques such as using a harness and swivel apparatus (see below). If the technique must be used, the investigator should familiarise the rat with the cage for a few days before it is forced to take up permanent residence in it.

The Bollman cage consists of two Perspex end plates approximately 16×8 cm which are screwed onto a base plate 16 cm apart. Eight 3 mm diameter steel rods with a 5 mm space between each rod are attached to both end plates to form the grid floor of the cage. Ten steel rods with screw threads at one end are arranged cylindrically above the floor, and pass through holes in one end plate and screw into holes in the second end plate. The internal diameter of this cylindrical arrangement is 5.5 cm. A second row of holes is made slightly interior to the first set on one side only so that the diameter of the cage can be reduced for smaller sized animals. Two centrally placed 2–3 cm holes in both end plates of the formed cage allow the insertion of a water bottle nozzle at one end and the exteriorisation of the tail at the other end. Food is simply supplied as pellets on the floor of the cage or with a slight modification of the anterior part of the floor, a small food container can be fitted underneath. One variation of this design of cage is to have the end plates 25 cm apart, and to introduce between them

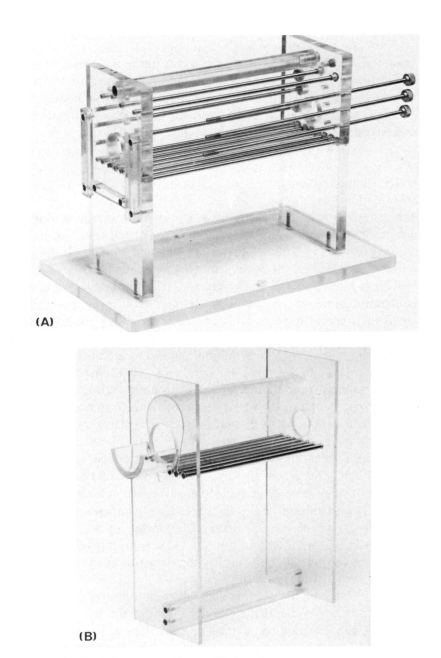

Fig. 103. (A) The Bollman cage, (B) a modified version.

an adjustable partition plate through which the steel rods pass. The plate can be moved backwards or forwards to allow the accommodation of animals of different lengths.

A modification of the basic design of the cage is seen in Fig. 103B. The cylindrical arrangement of steel rods has been replaced by a thin, acutely curved Perspex sheet, the ends of which fit through the slots in the floor. It is kept in place by adhesive tape wound completely around it and the floor of the cage. The Bollman cage can also be obtained commercially (see Equipment Index). To use the cage the steel rods on one side are slid back, and the rat, preferably still anaesthetised, is placed into the cage with its tail passing through the rear hole. The rods are screwed back into position so that they fit closely round the animal. Some forward and backward movement is allowed but the rat should not be able to turn back to front. If the rat is carrying an exteriorised catheter, this is passed out of the cage dorsally and as far to the rear of the animal as possible. The rear end plate hole should be used where the cage is closed by the curved Perspex sheet (see above). The catheter is taped to the outside of the cage, allowing a little slack for the movement of the rat. A complication that can arise consists of an unusual acrobatic manoeuvre by some rats in which they spin completely around their horizontal axis. This causes the catheter to wind round their body, and if enough slack has not been left the catheter becomes dislodged and useless. This manoeuvre can be prevented by taping down the rat's tail, again allowing some slack for movement.

17.3. Partial Restraint

To overcome the problem of the physiological disturbance to a rat confined in a Bollman-type cage, several procedures have been devised to allow the animal some freedom of movement while still offering protection of the catheter. Two such methods will now be described.

17.3.1. The Harness and Tether

This method depends upon the application of a harness round the rat and protection of the catheter with domestic curtain wire or a lightweight coiled spring which acts as the tether. One end of the tether is attached to the harness while the other is anchored to a swivel device which allows the rat almost complete freedom of movement while preventing the catheter from twisting and kinking. The basic steps involved in the use of the system are as follows. A blood vessel or other hollow organ is catheterised and the catheter is made to travel subcutaneously until it exits via a small incision at the dorsal nape of the neck (see 22.2). In a different application, a thermistor probe or other recording device can be conveniently placed within the body and also exteriorised via the neck. The harness is placed around the chest, incorporating the front legs and is closed round the body. The tether is attached by means of a saddle anchored to the harness while the catheter (or recording device leads) is passed up the tether and, in the case of the catheter, connected to the swivel. The catheter is now completely protected. The tether is either passed out of the cage and the swivel is clamped or the swivel is attached to the outside of the cage top. The component parts of the system are shown in Fig. 104 and the system in operation is shown in Fig. 105.

The swivel is the heart of the system and determines the complexity of the instrumentation. Swivels are available with a wide range of

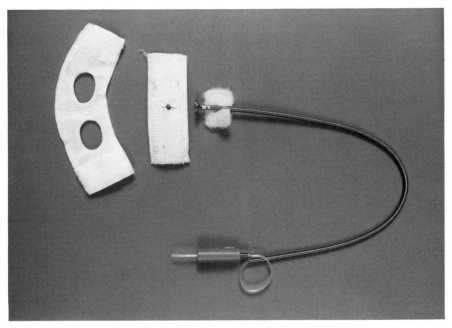

Fig. 104. Harness and swivel—component parts.

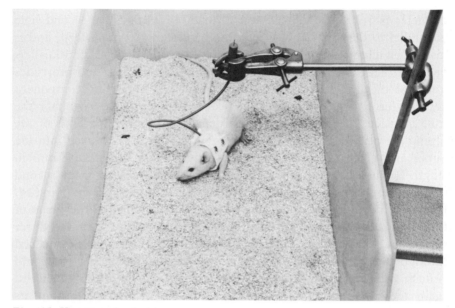

Fig. 105. Harness and swivel in use.

capabilities from a single fluid channel to four fluid channels and up to ten electrodes. The multi-channel system accommodates simultaneous drug infusion, blood pressure monitoring and sampling of body fluids in separate fluid channels. The electrical options allow the animal to be monitored by up to ten fully isolated electrodes. The harness and tether systems are in wide use and can be purchased commercially (see Equipment Index). However, these systems can be expensive or may in some instances be difficult to obtain and one that can be made easily in the laboratory has been described by Dalton *et al.* (1969).

17.3.2. *The Head Attachment Apparatus (adapted from Steffens, 1969)*

This method is designed to allow easily both sampling of fluids and the infusion of substances simultaneously. A Perspex cylinder is prepared, 8 mm in length with a diameter of 10 mm, in which are drilled two holes. One of these (Fig. 106A) coincides with the diagonal of the rectangular cross section containing the vertical axis of the cylinder. Its diameter is 1.2 mm, widened at both ends to 2 mm. The second hole is parallel to the vertical axis, 1.5 mm diameter and 4 mm deep, and is made at a site such that it does not meet the first hole (Fig. 106B). A piece of steel tubing 1.2 mm o.d. is fitted in the first hole and protrudes 6 mm at the upper end and 4 mm at the lower end. Two-component acrylic glue is placed into the widened ends of this hole to anchor the tube in place. After the glue has hardened, the protruding ends of the tubing are bent as in Fig. 106C and a polythene cap is plugged into the upper end of the tube. A piece of tubing 1.5 mm o.d. and 8 mm in length is glued into the second hole. From the lower end of the second hole a 1 mm diameter hole is drilled through the rest of the Perspex. From the lower aperture of this hole a groove parallel to the plane of the lower bent tube of the first hole is filed to the edge of the lower surface of the Perspex cylinder (Fig. 107A). A 21G needle is passed through the second hole and bent at right angles so that its lower end is flush with the outer end of the groove. This needle is now taken out again and a 9 mm length of 26G needle tubing is inserted through its hooked end. The needle is reinserted so that 4 mm protrudes freely and is then soldered in place (Fig. 107B, C).

The animal is now prepared to allow the Perspex attachment to be put in place on top of the head. To do this a median skin incision is made along the top of the head, the skin edges are retracted and the membranes are removed from the bones of the skull by rubbing with dry cotton wool. Three holes are made in the skull with a dental drill, one placed in the frontal bone and two in the parietal bone, forming a triangle just large enough to accommodate the Perspex cylinder. Care

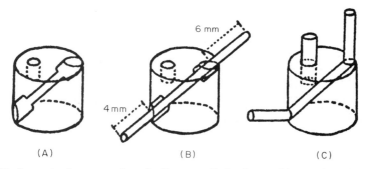

(A)   (B)   (C)

Fig. 106. Stages in the preparation of a Perspex cylinder for attaching to the rat's head.

must be taken not to puncture the underlying blood sinuses. Three stainless steel screws, 1.5 mm in diameter and 4.2 mm in length, are fitted into the holes but not tightened, so that about 0.5 mm remains between the skull and the head of the screws.

Catheterisation is now carried out. It is convenient to catheterise the large external jugular vein for sampling (see 22.2) and a small adjacent vein or the carotid artery for infusion. Both catheters are passed subcutaneously to emerge at the top of the head (see 22.2). The free ends of the catheters (which have been filled with a dilute solution of heparin-saline, 20 units heparin/ml saline, prior to the catheterisation) are slid over the lower ends of the two steel tubes of the prepared Perspex cylinder. The sampling catheter attaches to the first tube while the infusion catheter attaches to the tube in the second hole. The catheters should be cut down so that they are reasonably taut. It is necessary to measure the pieces cut off, and by knowing the original lengths of the catheters the final lengths can be gauged accurately. The dilute heparin solution is now replaced by one containing 500 units heparin/ml saline plus polyvinyl-pyrrolidine (molecular weight about 25 000) added to a total concentration of 40% w/v. The triangular area of the skull between the three screws is covered with a 1 mm thick layer of dental acrylic cement (see Equipment Index), and the Perspex cylinder is pushed firmly into this between the three screws. The cement that is squeezed out is arranged around the heads of the screws and the connections of the catheters with the steel tubing. When the cement has dried the skin incision is sutured around the base of the cylinder

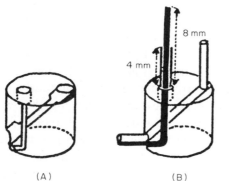
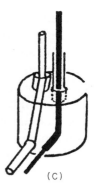

(A)   (B)   (C)

Fig. 107. Further stages of Fig. 106.

and its attachments. The rat is injected with benzylpenicillin (see Section 58) as a prophylactic measure.

The rat is now prepared for the attachments to the top of the fixed cylinder. For this, it can be placed in an open-topped high-walled cage. A flexible polythene tube, 1.5 mm i.d. and 3.5 mm o.d., is fixed to the steel tube in the second hole. A swivel can be interposed in this polythene tubing to allow greater freedom of movement of the rat if required. The tubing is taken outside the cage and fixed in such a way that it does not hamper the movements of the rat, for example by counterbalancing it with a light weight (see Fig. 108). The tube remains in place at all times.

To infuse substances, a thin polythene catheter 1.45 mm o.d. and 0.75 mm i.d. connected to an infusion pump is passed down the larger sized polythene tube connected to the second hole of the Perspex cylinder. To obtain blood samples, a long polythene tube 1.45 mm o.d. and 0.75 mm i.d. filled with saline and stoppered, is attached to the steel tubing at the top of the first hole. For use, a syringe is connected to the polythene tube and a few drops of blood are withdrawn to dislodge any blood clots. Next, 0.2 ml of saline is introduced to clear the tube. This is followed by 2 μl of dilute heparin-saline solution and then a small 2 mm air bubble is introduced briefly by opening the tube. More saline is injected until the air bubble is calculated, by knowing the total length of the polythene and steel tubing involved, to lie just above the end of the lower steel tube in the Perspex cylinder. A second air bubble is introduced which in turn is sent down the tubing with saline until the first air bubble is calculated to lie just above the entrance of the catheter into the blood vessel (Fig. 108) and separated from the circulation by the 2 μl of dilute heparin–saline solution. This air bubble prevents mixing of blood and saline and also acts as a marker. The following sampling procedure is then carried out. The contents of the polythene tube are withdrawn until blood just enters the nozzle of the syringe. A second syringe replaces the first and a blood sample is taken. If a second sample is required shortly after the first, the tube is cleared with 0.2 ml of saline, 2 μl of heparin are introduced together with an air bubble and the whole procedure just described is repeated. After sampling the tube is cleared of blood by injecting 0.2 ml of saline. This is followed by 0.2 ml of PVP–heparin solution until it fills the entire length of the catheter and tubing, using an air bubble as a marker. The sampling tube is removed and replaced by the polythene cap.

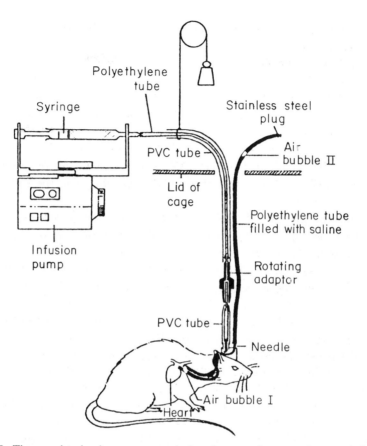

Fig. 108. The complete head apparatus attached to the rat and connected to two indwelling catheters and to an infusion apparatus (from Steffens, 1969).

If the method is to be used only for the sampling of blood or for infusion alone, then only the first hole is drilled right through the Perspex. The second hole is drilled for only 4 mm and is used solely for attaching the flexible tubing which is fixed outside the cage (see Fig. 106C). Although the method might seem rather elaborate, there is good evidence that once the rat is accustomed to the apparatus, for which a period of about 4 days must be allowed, any physiological signs of stress become absent and the determination of the true values of physiological parameters can be made with some confidence.

The same can be said for the harness and tether system of restraint and there is anecdotal evidence that such restrained animals will even engage in reproductive activity indicating the relatively stress-free nature of the restraint!

17.4. Care of the Restrained Animal

Because of the potential distress caused to the animal by restraint, the investigator should aim to minimise this by carrying out the following:

(1) Use the method of restraint which allows the animal the greatest freedom of movement which will still be compatible with achieving the aims of the procedure.
(2) Restraint should only be carried out by persons who are familiar with the procedure and the potential problems.
(3) The animal should be checked frequently during the day.
(4) Familiarise the animal to the restraint procedure for an hour or two for 3 or 4 days before the restraint.
(5) Restrain the animal for the shortest time possible.

Chapter Five

SPECIFIC SURGICAL OPERATIONS

18. INTRODUCTION

It is important that reference is made to Chapters Three and Four prior to attempting any of the surgical procedures described in this section. Irrespective of the type of surgical procedure, high standards of intraoperative and postoperative care are essential if meaningful experimental data is to be obtained. It is also important that investigators should gain sufficient experience in the use of surgical instruments to carry out procedures skilfully, with a minimum of trauma to the animal. Prior to attempting any procedure in a living animal, careful dissection of post-mortem specimens should be undertaken to ensure a thorough understanding of the relevant surgical anatomy.

19. ADRENALECTOMY

The two adrenal glands are small round pink-coloured organs, one situated near the anterior pole of each kidney and often loosely attached by means of adipose tissue.

For their removal, the anaesthetised rat is placed on its ventral surface with its tail towards the investigator. No restraint of the animal is required. In this position the back of the rat naturally assumes a slight humped-back posture. A small 1–2 cm midline incision is made in the skin ending just behind the peak of the hump (Fig. 109). Scissors are inserted subcutaneously through the incision and pushed alternately down either side of the animal for a short distance and opened several times to blunt dissect the connective tissue to allow access to the muscle. The skin incision can then be retracted with forceps, first to one side to remove one gland and then to the other side to remove the second gland. Entrance into the peritoneal cavity on the left side, just posterior to the last rib, is gained by making a small cut with pointed scissors into the muscle a few millimetres below the spinal muscles, i.e. about one-third of the way down the side of the rat (Fig. 110). The muscle incision should be made only just wide enough to allow the passage of the adrenal gland. In practice accurate control of the width of the incision is obtained by first scoring, but not cutting right through, the muscle with scissors, and then forcing the points of a pair of forceps through the remaining muscle layers. The width of the hole so produced is regulated by expanding the scissors or forceps. The correct placing of the incision should allow the anterior portion of the spleen to be seen

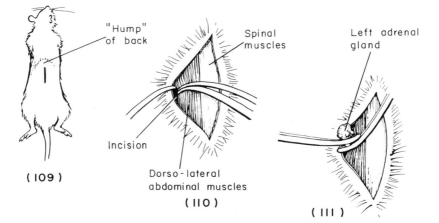

Fig. 109. Site of incision for adrenalectomy. *Fig. 110.* Entrance into the abdominal cavity. *Fig. 111.* Removal of the left adrenal gland, the two pairs of forceps are pulled away from each other.

directly underneath or slightly to the left. A pair of curved forceps held in the left hand are inserted into the peritoneal cavity so that the spleen is made to lie lateral to the concave side of the forceps and is prevented from moving medially and obscuring the gland. A second pair of forceps are inserted deeply, and the incision, together with both pairs of forceps, is moved medially. This manoeuvre is designed to push the greater part of any loose fat out of the way and to bring the incision over the gland. Some deep searching for the gland may be necessary if much fat is present since the fat tends to cause the gland to lie ventrally and may also totally obscure it. A sighting of the kidney will help pinpoint the position of the gland. Once found, the gland is pulled out through the muscle incision by holding the periadrenal fat. On no account must the gland itself be grasped or a piece may become detached and will remain to reimplant and function in the abdominal cavity.

Removal of the gland is effected by clamping the tissue and blood vessels at its base with both pairs of forceps as in Fig. 111. The forceps are then drawn apart sharply from each other causing the gland to tear away. Any fatty tissue is returned into the abdominal cavity. No other attempt at haemostasis is necessary as any bleeding is small and inconsequential.

On the right side, the adrenal gland and the anterior part of the kidney are overlain by the peripheral part of the liver. After the muscle incision, which should be made about one-quarter the way down this side of the rat, the liver is pushed away anteriorly by the convex surface of one pair of forceps while the adrenal gland is withdrawn with the second pair. The process of removal of the right gland is identical to that for the left gland. In immature rats little or no fat will be present to obscure the adrenal glands which lie superficially and are thus very easily seen. No suturing of the muscle incision is necessary providing it has been kept small. However, in old or fat rats, a larger incision may have to be made to facilitate finding the glands, and such an incision will require a single suture to prevent herniation.

If no postoperative treatment is given to prevent or replace the loss of sodium which occurs as a result of the operation, then adrenalectomised rats will survive only for between 7 and 28 days, depending on the strain and age of animal. Long-term survival is effected either by the administration of corticosteroids (e.g. hydrocortisone, see 57) or more usually by giving the animals a saline/sucrose (0.9% saline, 10 g/l sucrose) solution to drink *ad libitum*.

19.1. Adrenal Demedullation (adapted from Griffith and Farris, 1942)

The adrenal gland is brought out of the abdominal cavity as above, and the fat and blood vessels at its base are clamped with forceps. The distal tip of the gland is clipped away with sharp scissors. Fine smooth-tipped curved forceps are closed gently, first about the base of the gland itself applying pressure, and then moved progressively towards the open end of the adrenal capsule. The body of the gland pops easily out of the capsule, leaving the capsule plus a layer of glomerulosa cells. Enough cortical tissue regenerates from these remnants of the gland to meet the physiological requirements for the cortical hormones. Postoperatively the rats are given saline/sucrose to drink (see above) until regeneration of enough cortical tissue is obtained, usually within 5 weeks.

20. BILE DUCT CATHETERISATION

The bile duct in the rat is about 1 mm wide, is rather translucent, and runs from the hilum of the liver where it collects tributaries from the liver lobes, through the pancreatic tissue, to the duodenum into the lower end of which it discharges its contents. There is no gall bladder in the rat. The bile duct also serves as the main duct for the collection of the pancreatic secretions. Catheterisation of the duct near the hilum of the liver will allow the collection of pure bile, while catheterisation in the pancreatic region with appropriately placed ligatures will result in the collection of the pancreatic juices (see Fig. 112).

20.1. Collection of Bile

The anaesthetised rat is placed on its back with its tail towards the investigator. A midline abdominal incision is made for about half the length of the abdomen posterior to the xiphoid cartilage. A small bolster prepared from gauze or a large plastic tube is placed under the thorax to arch the back and facilitate the exposure of the duct. A large-bore needle is passed through the skin at the side and towards the back of the rat, and into the abdominal cavity taking care not to puncture the gut (Fig. 113). The needle will be used to exteriorise the catheter and its internal diameter should be just large enough for the catheter to pass through.

The duodenum and a small part of the intestine is pulled out to the right of the investigator and placed on a gauze pad, moistened with physiological saline, on the abdomen. The course of the bile duct amidst the pancreatic tissue can now be seen, especially if it is traced back from the hilum of the liver. The major lobes of the liver are pushed back out of the way against the diaphragm, but if they show a reluctance to remain there of their own accord a retractor can be used or the lobes can be pulled out and placed on the chest wall after being wrapped in a saline-moistened gauze pack. Subsequent procedures are facilitated by the use of a binocular dissecting microscope.

Since about 0.5–1 cm of duct near the hilum of the liver is free of pancreatic tissue, this part must be used for the catheterisation. Forceps are carefully pushed through the connective tissue under this part of the duct and opened to free this length of bile duct from connective tissue. A small length of doubled up thread is pulled through under the duct and cut to obtain two threads to form two ligatures. One ligature is tied tightly immediately in front of the pancreatic tissue to prevent the flow of bile. The obstructed duct becomes turgid but does not dilate. The second ligature is partly tied (first half-hitch) but left loose, 4 or 5 mm from the first ligature and near to the liver (Fig. 114). A polyethylene catheter, 0.28 mm i.d. and 0.61 mm o.d., is cut so that a short bevelled point is produced. The bile duct near to the first ligature is held with fine forceps (e.g. watchmaker's forceps) (Fig. 81) and partially transected with fine scissors, and the catheter is pushed into the duct towards the liver and past the loose second ligature. This ligature is now tightened and tied to make a double reef knot to secure the catheter in the bile duct. The free ends of the first ligature can be also used to secure the catheter. Bile should be seen immediately flowing in the catheter.

The end of the catheter is pushed through the large-bore needle to bring it out of the abdomen, and the needle is removed (Fig. 115). When exteriorising the catheter, care must be taken not to allow it to twist. It should lie naturally otherwise the bile duct will twist and the flow of bile will stop. The gut is returned into the abdominal cavity on top of the catheter and the flow of bile is again checked. If satisfactory, the catheter is secured to the skin where it emerges through the skin

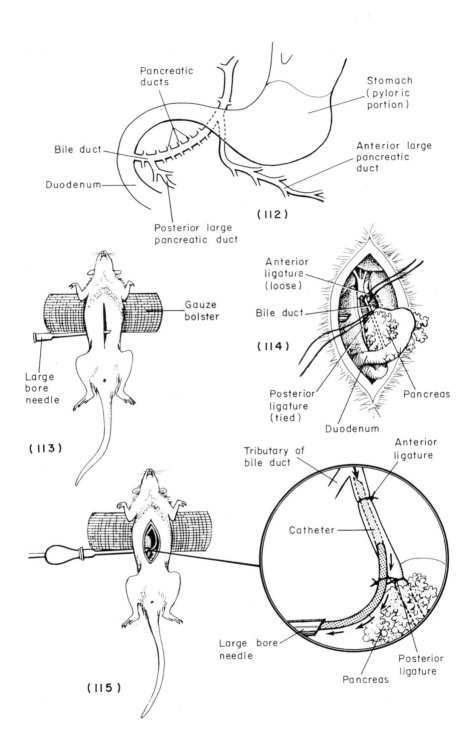

Pancreatic ducts

Stomach (pyloric portion)

Bile duct

Anterior large pancreatic duct

Duodenum

(112)

Posterior large pancreatic duct

Gauze bolster

Large bore needle

(113)

Anterior ligature (loose)

Bile duct

(114)

Posterior ligature (tied)

Duodenum

Pancreas

Tributary of bile duct

Anterior ligature

Catheter

(115)

Large bore needle

Posterior ligature

Pancreas

by a single suture and the abdominal incision is closed. The rat is placed in a restraining device (e.g. a Bollman cage) so that the catheter can be protected and the bile collected when the animal regains consciousness (see 17.2). If the intention is to collect bile for an extended period then restraint of the animal and protection of the catheter using a harness and tether (see 17.3.1) should be employed. The end of the catheter should be below the level of the animal during bile collection to aid flow by gravity. For drug metabolism studies, a system such as that described by Tomlinson *et al.* (1981) in which the rat is maintained in a metabolism cage works very well.

Apart from the need to restrain the animal to prevent it removing the catheter, no other special measures are required other than the provision of a high standard of general postoperative care. The production of bile should be about 10–15 ml in 24 hours. Bile can be collected for up to several weeks if desired, the main regulatory factor being the type of restraint placed on the animal.

20.1.1. Intermittent Collection of Bile with Recirculation

In some instances, it is desirable to collect bile intermittently over a period of days or longer without significant interference to the enterohepatic circulation. This can be achieved by using a re-entrant catheter placed in the bile duct and connected to a suitable sampling point attached externally on the body. The method has been described by Chipman and Cropper (1977).

20.2. Collection of Pancreatic Secretions

The surgical approach is exactly the same as described in 20.1 for the collection of bile. A ligature is tied around the bile duct where it enters the duodenum and a second one in the hilum of the liver before the duct bifurcates. A catheter placed in the duct near to either of these ligatures will collect pure pancreatic secretion. The pancreas itself must be handled as little as possible. The catheter is exteriorised and the

---

Fig. 112. The bile duct and its anatomical relationships (from Lambert, 1965). *Fig. 113.* Preparation of the rat for bile catheterisation. *Fig. 114.* Placing the two ligatures round the bile duct. *Fig. 115.* The exteriorised catheter in the bile duct.

animal restrained as described for the collection of bile (see above). No special postoperative therapy is necessary but a 5% glucose drink will help the flow of secretion. Rats will survive for up to a few weeks, but the biliary obstruction which causes jaundice is detrimental to prolonged survival. The flow rate of pancreatic secretion is variable but usually 0.5–1 ml/h.

20.3. Simultaneous Separate Collection of Pancreatic and Biliary Secretions

The bile duct is ligated at its entrance into the duodenum. The duct is then partially transected about 1 cm from the hilum of the liver, just before the start of the pancreatic tissue. One catheter that collects bile is passed into the duct a short way towards the liver and ligated. A second catheter is passed into the duct through the same cut, if the duct is wide enough to admit it, a short way towards the duodenum and ligated. This collects pancreatic secretion. Alternatively, the second catheter can be passed into the duct through a cut made near the duodenal end of the duct. The two catheters are exteriorised through the flank skin. Rats can survive for several weeks with combined pancreatic and biliary catheters.

21. CAESARIAN SECTION

This operation accomplishes the removal of rat foetuses from one or both uterine horns, usually just before parturition. It can be performed, however, at any time after the reasonable development of the foetus (about day 13 onwards if the foetus is not required to live or about day 20 to obtain viable foetuses).

The pregnant animal is laid on its back with its tail towards the investigator. Entrance is gained into the peritoneal cavity through a full-length abdominal skin and muscle incision (Fig. 116). One uterine horn is pulled out and placed on a warm saline-moistened gauze pack. The horn is slit along its whole length with scissors, along the side opposite

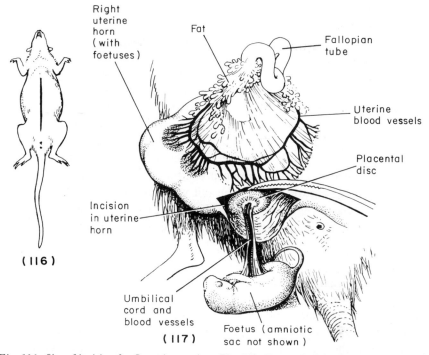

Fig. 116. Site of incision for Caesarian section. *Fig. 117*. Removal of the foetus, the placental disc is pulled away from the uterine wall.

to the placenta and blood vessels. If only one or two foetuses are to be removed then the slit is made large enough to allow only the selected foetus to be removed. If pregnancy is well advanced, the blood vessels to each placental disc will be large and may need to be ligated. In the earlier stages of pregnancy there is no need to ligate the blood vessels as bleeding is slight. Curved forceps are placed under the disc-shaped placenta which is then pulled away from its connection with the uterus. In this way the complete foetus in its amniotic sac and attached to the placenta is removed (Fig. 117). Bleeding at the point of attachment of the placenta to the uterus is stopped with pressure from cotton-tipped applicators. If the rat is intended to recover from the procedure, the uterine horn should be sutured using 4/0 catgut, and returned to the abdominal cavity. The abdominal muscle and skin incision are then closed.

22. CHRONIC CATHETERISATION OF BLOOD VESSELS

It is frequently necessary to catheterise blood vessels in order to administer substances or withdraw blood samples over a prolonged period of time, or to make direct blood pressure recordings (Fig. 118). In most cases, the catheter will need to be exteriorised and protected from the attentions of the animal, who will have to be restrained appropriately (see 17). Most blood vessels which are not too small and are reasonably accessible can be catheterised. The method will be illustrated, however, by describing procedures for catheterisation of the jugular vein and carotid artery, the tail vein, and the dorsal aorta both directly and via the femoral artery. The use of vascular access ports as an extension to the catheterisation technique will be described.

22.1. General Considerations

There are three major considerations when requiring to catheterise a blood vessel. These are the type of catheter to use, the prevention of infection and the prevention of clotting within the catheter, which would obstruct flow.

22.1.1. Type of Catheter

The ideal characteristics of a catheter material are thromboresistance, ease of insertion, long-term tolerance, infection control and favourable mechanical and convenience factors. Catheters can be obtained in a variety of materials, such as polyvinylchloride, polyethylene, polyurethane, silicone rubber and polytetrafluoroethylene. Each has advantages and disadvantages (see Gay and Heavner, 1986). However, the most frequently used are polyethylene and silicone rubber. Polyethylene (e.g. Portex[R]) has good flexibility but is rigid enough to handle easily and can easily be moulded in hot water. Its main disadvantage is that it cannot be autoclaved for sterilisation and it may become brittle and break if left under the skin for long periods. Silicone rubber (e.g. Silastic[TM], Silasol[R]), is extremely flexible which makes it more difficult to work with, however less infection is encountered with its use and it

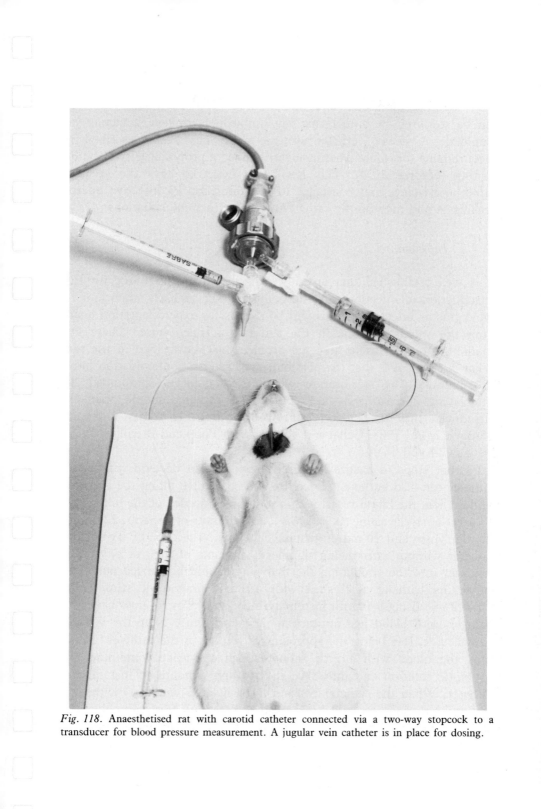

Fig. 118. Anaesthetised rat with carotid catheter connected via a two-way stopcock to a transducer for blood pressure measurement. A jugular vein catheter is in place for dosing.

is less likely to abrade the endothelial lining and cause inflammation. It can be sterilised by autoclaving but unlike polyethylene it cannot be moulded. A newer catheter system which has been described is polyurethane to which is permanently bonded a proprietary hydrophilic coating (Hydrocath™). This has been shown to have the greatest thromboresistance and resistance to contamination by infective micro-organisms and may be the catheter of choice in some situations.

22.1.2. Infection

It is well established that an indwelling catheter is a source for the introduction of infection into the body. It is particularly important to avoid this as infected animals will be useless for experimentation apart from the welfare aspects in the suffering it may cause. There are conflicting reports as to how the problem may be tackled in the rat. What is essential is that the possibility of bacterial contamination must be minimised by sterilisation with disinfectants or eliminated by autoclaving the catheter (see 15.4). The catheter should then be introduced into the blood vessel using reasonable aseptic methods (Popovic *et al.*, 1963; Gellai and Valtin, 1979; Popp and Brennan, 1981; see also 14 and 15).

Further steps to ensure that infection does not develop within the catheter can also be taken if desired. These involve filling the catheter while it is in the blood vessel with a solution composed of the following. To 1 ml of sterile saline (0.9%) (w/v), add 20 units of heparin, 225 units chymotrypsin and 20 mg of gentamycin (Palm *et al.*, 1991). To fill the catheter without introducing significant amounts of this, or any other solution, into the circulation the following should be carried out. The indwelling catheter is flushed with a small amount of saline (e.g. 0.1–0.2 ml) using a 0.5 ml insulin syringe. Very slowly draw back the plunger until blood just appears in the needle hub. Note the volume drawn back, this being the approximate volume of the catheter. Flush back the blood with a little saline, attach a syringe containing the antibiotic solution and introduce the required amount to just fill the catheter. When the catheter is again to be flushed (see later) withdraw the antibiotic solution and discard and replace with fresh after flushing back the blood with saline. At no time should significant amounts of heparin be introduced into the circulation as it may produce unwanted bleeding. Instead of preventing infection by filling the catheter with an antibiotic solution, a different approach may be used whereby the antibiotic is bonded to the catheter material, though this has certain limitations (see below).

22.1.2. Catheter Obstruction

Synthetic catheters cause responses in the blood which tend to induce deposition of a platelet–fibrin complex (thrombus) in the tip of the catheter which blocks it. This is particularly so if the endothelial lining of the blood vessel is abraded by excessive manipulation during implantation of the catheter. The problem, if serious, can be tackled by using one or both aspects of a two-pronged approach.

(1) The catheter is soaked for 30 minutes in 5% tridodecyl-methylammonium chloride (TDMAC) in 95% ethanol (w/v) (see Equipment Index). It is then air dried and washed five times with distilled water. Following this, it is filled with heparin solution, 100 units/ml, and incubated for 30 minutes. It is again allowed to air dry and washed five times with distilled water. The prepared catheter can be disinfected before use. The bonding technique can also be carried out using a TDMAC–heparin complex purchased commercially (see Equipment Index) and the directions given by the manufacturer should be followed. The technique of bonding can also be used to bond antibiotics to prevent catheter infection. In this case the TDMAC pretreated and washed catheter is incubated for 30 minutes with a penicillin solution (10 mg/ml) before being air dried and washed (Trooskin et al., 1983). It should be noted that with frequent flushing, the action of the heparin and antibiotic will be reduced as they become washed off the surface with time.

(2) If the catheter becomes blocked, it may be possible to dissolve the thrombotic occlusion by filling the catheter with a solution of urokinase or streptokinase (Hurtubise et al., 1980). This should be left for 10 minutes and then withdrawn and the catheter flushed with sterile saline before being filled with a heparin–saline solution (20 units heparin/ml). The procedure can be repeated if not initially successful.

22.2. Catheterisation of the Jugular Vein

The anaesthetised rat is laid on its back with its head towards the operator (Fig. 119). The jugular vein (the right external jugular vein

Fig. 119. Site of incision for right jugular vein catheterisation.

is most often used) is exposed as for i.v. injection (see 2.15.2). At least 1 cm of vein, up to the point where it disappears under the pectoral muscle, should be *gently* cleaned of fat and other tissue by blunt dissection. A pair of forceps is passed under the vein and a piece of doubled up thread is withdrawn and cut to give two pieces. The anterior thread is tied tightly to occlude the vein. Alternatively, occlusion can be carried out using a ligating staple (see Fig. 83). The first half-hitch of the posterior thread is formed but left loose several millimetres posterior to the first ligature (Fig. 120). The vein is then placed under light tension either by spreading a pair of iris forceps under it or by raising the ends of the posterior ligature vertically. This may require an assistant. The vein is then semi-transected with fine scissors between the two ligatures. A catheter, i.d. about 0.5 mm and o.d. 0.61–1.00 mm,

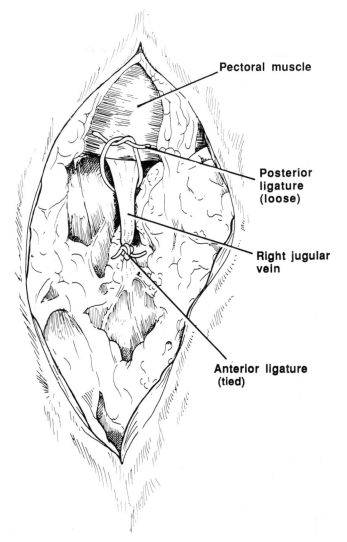

Fig. 120. Placement of ligatures.

is prepared by cutting squarely across it with a scalpel blade to produce a blunt end (the end should preferably not be bevelled). The tip of the catheter is inserted into the vein and pushed forwards towards the heart, this being facilitated either by opening the incision with fine forceps (Fig. 81) or by inserting a disposable catheter introducer (see Equipment

Index). The posterior ligature is then tied round the vein and catheter (Fig. 121). If the catheter is to be used only for administering substances then only a few millimetres need to be within the blood vessel. For blood withdrawal or blood pressure measurement it is important that the catheter is pushed in until it lies either within the superior vena cava or at the entrance to the right atrium (Fig. 122). The distance the

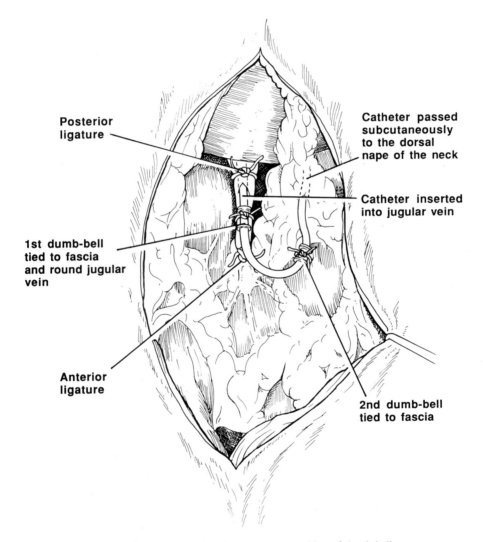

Posterior
ligature

Catheter passed
subcutaneously
to the dorsal
nape of the neck

Catheter inserted
into jugular vein

1st dumb-bell
tied to fascia
and round jugular
vein

Anterior
ligature

2nd dumb-bell
tied to fascia

Fig. 121. Placement of catheter into jugular vein—note position of dumb-bells.

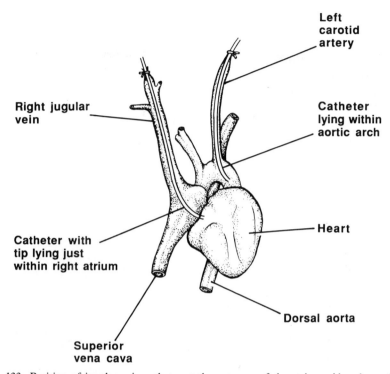

Fig. 122. Position of jugular vein catheter at the entrance of the atrium. Also shown is the position of a carotid artery catheter within the descending limb of the aortic arch.

catheter has to be pushed forward will have to be measured in preliminary experiments. This distance is fairly constant for a given body weight. For example, for the Sprague–Dawley rat in a body weight range of 300–500 g, between 3.0 and 4.1 cm of catheter need to be inserted. The correct positioning of the catheter is tested by gauging the ease with which blood can be withdrawn after attaching a needle and syringe filled with saline. If satisfactory, the blood should be flushed back into the circulation. After flushing with a small amount of saline ensuring no air gets into the circulation, the catheter is finally filled with a heparin–saline solution (20 units heparin/ml) ("heparin lock") and a suitable stainless steel pin inserted into the end of the catheter. This pin can be fashioned from a suitably sized hypodermic needle which has been separated from the hub and filled with silicone rubber cement (Silcoset, see Equipment Index). The catheter is then tied into the blood vessel with the posterior ligature. If thread is used to form the anterior ligature, the ends of this can be used to further secure the catheter.

To exteriorise the catheter, a small incision is made in the dorsal nape of the neck after first removing the hair and swabbing the skin with antiseptic. A pair of straight forceps or a trocar of 16G or larger (see Equipment Index) is passed subcutaneously from the dorsal incision down the side of the neck to emerge anterior to the site of the entry of the catheter into the jugular vein. The end of the catheter is grasped and withdrawn subcutaneously or is passed subcutaneously through the trocar needle. However, just prior to this manoeuvre, it is recommended that a loop of catheter ("stress loop") is prepared either within the area of the neck incision or in a subcutaneous pocket near to the dorsal exit site. This loop provides slack catheter which allows for the animal's normal body movements. Without it there is a danger of the catheter tip being pulled back or out of the vessel. Alternatively, where the catheter has been securely fastened to the fascia using dumb-bells or retention cuffs (see below), a stress loop may not be necessary. After exiting, the catheter is then either cut so that about 25 mm remains protruding and stoppered with a pin, or the animal is restrained and the catheter is passed outside the holding cage (see 17.2). Finally, the skin incisions are closed.

A number of points should be noted for jugular vein catheterisation and some will be applicable to catheterisation of other blood vessels. First, the vein sometimes narrows considerably on handling and this should, therefore, be kept to the minimum. Applying a few drops of local anaesthetic (e.g. lignocaine, without adrenaline) to the vein prior to handling may prevent constriction, but in any case leaving the vein untouched for a few minutes usually allows it to regain some of its former size. Second, prior to insertion of the catheter into the blood vessel, it should be filled with a solution of heparin in saline (20 units/ml), or other suitable heparin-containing solution (see 22.1) to prevent clotting and the end stoppered with a pin to prevent the solution from escaping ("heparin lock"). A solution of 40% polyvinylpyrrolidone (PVP, molecular weight about 25 000) in saline (v/v) containing 50 units heparin/ml might also be found useful as the viscous PVP prevents diffusion of blood into the catheter. Only enough heparin solution to fill the catheter must be used. In spite of the "heparin lock", clotting within the catheter can occur. To prevent this a maintenance schedule is required whereby the heparin solution is removed, usually daily, and the catheter refilled with fresh heparin–saline after first returning the blood into the circulation and flushing any residual blood out of the catheter with 0.1–0.2 ml sterile saline. Third, when using polyethylene or polyvinylchloride catheters, they may become kinked where they bend in the neck before passing to exit dorsally. One means of preventing this is to fashion two small dumb-bells out of a polyethylene catheter with an internal diameter the same as or fractionally larger than the external diameter of the catheter to be implanted. For example, for a catheter of o.d. 0.86 mm (Portex ref. 800/100/160) dumb-bells are made from a catheter of i.d. 1.02 mm (Portex ref. 800/100/280). The dumb-bells are approximately 3 mm in length and made by flaring both ends by flaming or using a hot soldering iron. The dumb-bells are then soaked for a few minutes in xylene which causes them to swell. They are then washed in sterile water and can be slipped easily over the catheter. One dumb-bell is passed over the catheter to come to rest close to where the catheter will emerge from the blood vessel. Therefore, it has the useful function of identifying how much catheter has to be inserted into the blood vessel. The second dumb-bell is positioned approximately 25 mm from the first but this may need to be varied depending on circumstances. After a short while the dumb-bells contract to their original size and adhere tightly to the catheter. Alternatively, dumb-bells can be glued to the catheter using special cyanoacrylate cement (see Equipment Index). The dumb-bells are used for tying the catheter to the adjacent fascia and maintaining a gentle bend in the catheter to prevent it kinking (Figs 121 and 123, also see Gellai and Valtin, 1979).

Although kinking is not a problem when using silicon rubber tubing, difficulties arise when attempting to ligate it, since the lumen can easily be narrowed and thus obstruct flow. This can be overcome in several ways. Small 2–3 mm silicone rubber retention cuffs can be placed over the implanted catheter in exactly the same way as for the dumb-bells above (Fig. 123) or left so that they can be moved up and down the catheter. The cuffs can be glued in place (Silastic glue—see Equipment Index). Catheters containing these cuffs can also be purchased commercially. Ligating gently behind each cuff will prevent movement in the catheter. Secondly, by cutting two holes in a flange of silicone rubber sheeting, about 3 mm × 6 mm, and threading the catheter through, the flange can be suitably placed near to the exit of the catheter from the blood vessel and can be used to secure the catheter to the surrounding fascia. Again the flange and the catheter can be glued together (Garner et al., 1988). Thirdly, a silicone rubber catheter emerging from the

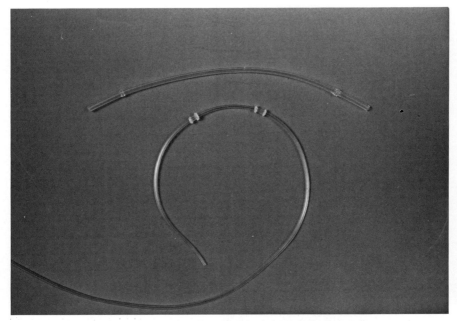

Fig. 123. Catheters showing retention rings (top) and dumb-bells (bottom).

blood vessel can be stretched over the flared end of a suitably sized polyethylene catheter and tied in place. Securing ligatures can then be placed round the more rigid polyethylene catheter before it is passed subcutaneously to the dorsal neck exit site. This "mixed" catheter can be prepared prior to surgery. A variation on this approach is described below.

22.3. Catheterisation of the Carotid Artery

The carotid artery lies medial to and below the jugular vein and is exposed in a similar manner to the latter by blunt dissection between the sternohyoid, sternomastoid and omohyoid muscles (Figs 124 and 125). Care should be taken that the adjacent vagus nerve is dissected away from the artery. Because subsequent passage of a catheter along the right carotid artery may result in it emerging in the left ventricle and causing arrhythmias, the left carotid is generally employed. A catheter in the artery on this side will pass into the aortic arch with no ensuing problems. An occluding ligature (thread or staple) is placed on

Fig. 124. Site of incision for left carotid artery catheterisation.

the artery as anteriorly as possible and a loose posterior ligature is positioned several millimetres away (Fig. 126). It may be found helpful to stabilise the vessel by positioning a pair of artery forceps or plain dissecting forceps under the vessel. A small clamp (e.g. Scoville Lewis clip) is placed on the most distal portion of the carotid artery to stop the flow of blood. A small incision is made using microscissors only large enough to allow a suitable sized catheter (usually polyethylene or other with equivalent rigidity) to be forced through. The posterior ligature is lightly tightened round the artery and catheter and the catheter is pushed towards the heart while the clamp is released, until the tip (blunt-ended) lies in the aortic arch, most often so that it lies downstream in the descending portion (see Fig. 122). This distance is determined previously. The posterior ligature is then tied round the

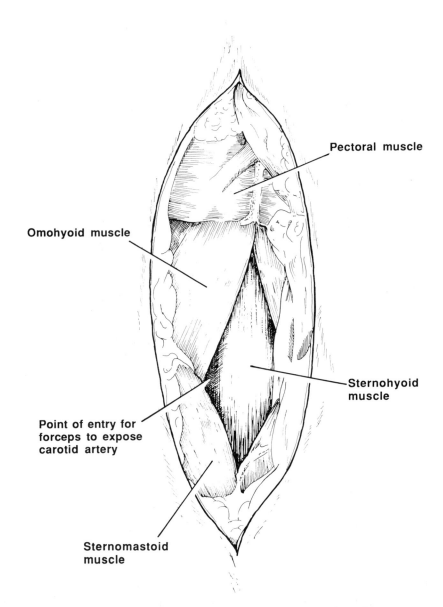

Pectoral muscle

Omohyoid muscle

Sternohyoid muscle

Point of entry for forceps to expose carotid artery

Sternomastoid muscle

Fig. 125. Arrangement of muscles and point of entry for exposure of carotid artery.

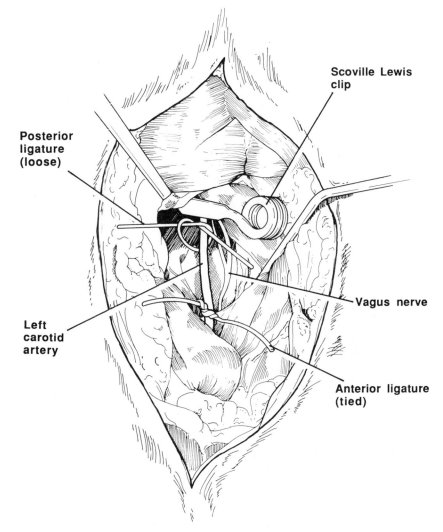

Scoville Lewis
clip

Posterior
ligature
(loose)

Vagus nerve

Left
carotid
artery

Anterior ligature
(tied)

Fig. 126. Placement of ligatures and clamp.

artery and catheter and the catheter can be further secured by the free
ends of the anterior ligature (Fig. 127). The catheter is then exteriorised
at the dorsal nape of the neck (see above). Points noted for jugular vein
catheterisation are mostly applicable for catheterisation of the carotid
artery. It is self-evident that insertion of apparatus other than catheters

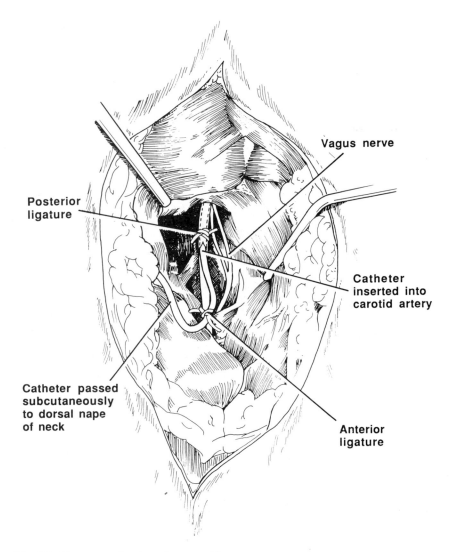

Vagus nerve

Posterior
ligature

Catheter
inserted into
carotid artery

Catheter passed
subcutaneously
to dorsal nape
of neck

Anterior
ligature

Fig. 127. Placement of catheter into carotid artery.

can be made in blood vessels. For example, insertion of a thermocouple probe (0.5 mm o.d.) would allow cardiac output measurements to be made by a thermodilution technique when combined with catheterisation of the jugular vein (Hanwell and Linzell, 1972).

22.4. Catheterisation of the Tail Vein (adapted from Born and Moller, 1974)

To carry out this method it is necessary to have 2.5 cm wide adhesive tape, a 0.32 cm i.d. compression spring, and electrical shielding 0.79 cm flat width which is made of braided tinned copper. A small "trapdoor" flap is made over one vein about 5 cm from the base of the tail by cutting on either side of the vein with a scalpel and across one end, taking great care not to cut the vein which lies immediately beneath the skin. The flap of skin is pulled back and the vein exposed and cleaned of connective tissue by brushing with cotton gauze. The vein is dilated by applying a tourniquet (a piece of string) round the base of the tail. The vein can then be entered either by partly transecting it with scissors and pushing in a polyethylene catheter, i.d. 0.28 mm, o.d. 0.61 mm, or by piercing the vein with a 20G needle and pushing a catheter, i.d. 0.28 mm, o.d. 0.61 mm, through the needle and then withdrawing the needle leaving the catheter in the vein (Fig. 128) (see also 2.15.1.2). About 1.5 cm of catheter should lie in the vein. The catheter should be filled with the infusion or injection material prior to the insertion, or filled with PVP–heparin solution (see 22.2) if intermittent injections are to be carried out. The tourniquet is released after insertion of the catheter. To verify that the catheter is truly i.v. a little of the injection material should be injected. To anchor the catheter in place a 1.25 ×

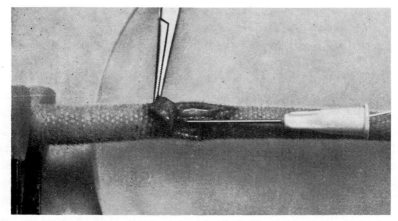

Fig. 128. Tail vein catheterisation. A "trapdoor" of skin is formed, a caudal vein exposed and a 20G needle with the catheter is inserted.

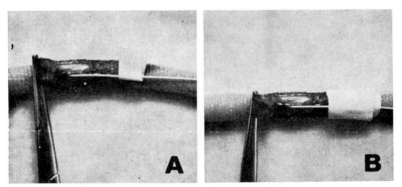

Fig. 129. (A) The adhesive tape is first wrapped around the catheter to form a flag; (B) the catheter is secured by a strip of tape wound around the "flag" and the tail.

2 cm piece of adhesive tape is wound first around the catheter then around the tail distal to the skin flap. This prevents the catheter sitting directly on the skin. A second piece of tape is wound around the first piece (Fig. 129), and the skin flap is replaced and kept in place by several turns of adhesive tape.

To protect the point of insertion into the vein and the external portion of the catheter, a flexible sheath is prepared. The tail is placed lengthwise along a double 12.5 × 2.5 cm strip of tape, the two tapes being stuck together by their adhesive sides, i.e. the tail is placed on the non-adhesive surface. This strip of tape is then folded round the tail to form a hollow tube, and tape is wound round it to keep it in place. The hollow tubes reduce the amount of adhesive that comes in contact with the tail and therefore reduce irritation. A 30 cm section of compression spring is threaded over the catheter and taped to the inside of the distal end of the hollow tube before it is finally closed (Fig. 130); this spring prevents the rat from chewing the catheter. A 12.5 cm length of electrical shielding is slid over the compression spring and the hollow tape tube and secured with tape proximally and distally (Fig. 131). The rat is placed in a flat-topped cage and the compression spring and catheter are passed through the top of the cage, the catheter is connected to an infusion apparatus or is stoppered until required for use. A short cross-rod is wired to the last few centimetres of compression spring on top of the cage so that the spring and tail are upright (Fig. 132). The rod

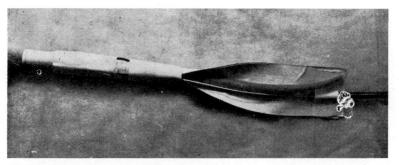

Fig. 130. The long edges of the flexible sheath have been partially taped together and the leading end of the compression spring anchored to the distal end of the sheath.

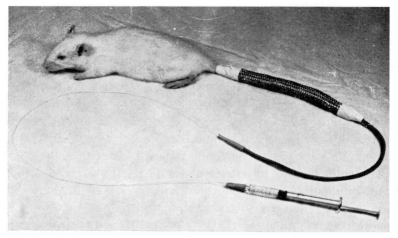

Fig. 131. The electrical shielding has been secured with tape and the rat is now ready for infusion.

prevents the catheter from being pulled back into the cage and allows the animal 360° mobility within the cage, thereby reducing the problem of restraint stress which is a consequence of some other forms of restraining and protection apparatus.

Fig. 132. The rat is shown in a flat-topped cage, connected to the infusion assembly (from Born and Moller, 1974).

22.5. Catheterisation of the Dorsal Aorta (and Femoral Artery)

22.5.1. Direct Catheterisation

The abdomen is opened and the dorsal aorta is located posteriorly where it bifurcates into the two common iliac arteries. The thin peritoneal covering and any connective tissue are removed by blunt dissection. A soft clamp or atraumatic artery forceps are used to minimise the problem of aortic aneurysm and these are placed on the aorta about 2.5–3 cm above the bifurcation. A small hole is made in the aorta with microscissors (Fig. 79) or a straight-edged cutting needle just above the bifurcation, and a polyethylene catheter is inserted to fit tightly and pushed anteriorly for a distance of about 2 cm. The o.d. of the catheter should be about 25% of the cross-sectional area of the lumen of the aorta. This degree of reduction of the cross-sectional area has no deleterious effect on hind limb function. The catheter is then anchored to the deep back muscles where it emerges from the aorta with a single ligature. Note that no ligature is placed round the aorta but a drop of tissue adhesive can be placed at the point of insertion of the catheter to secure it. The clamp is removed. The catheter is then made to follow the pelvic floor and brought out through the abdomen. The abdominal muscle incision is closed and the catheter is anchored to the abdominal muscle at the posterior end of the incision. The catheter is then carried under the skin to emerge at the dorsal base of the neck (see 22.2) and the skin incision is closed. Note that the catheter must be filled with a dilute heparin–saline solution or with a PVP–heparin solution and stoppered prior to its insertion into the aorta (see 22.2). After emerging from the neck, the catheter, which should be cut to protrude by about 1–2 cm in length, is stoppered by inserting a suitable stainless steel pin which can be removed each time the catheter is used. Exteriorisation using a harness and tether can also be carried out depending on requirements (see 17.3).

22.5.2. Indirect Catheterisation via the Femoral Artery

A catheter can be placed in the dorsal aorta by first catheterising the femoral artery and sliding the catheter into the aorta. This allows greater security to be achieved as the catheter can be tied to the femoral artery. For best results a "mixed" catheter is used and prepared as follows (Fig. 133). An 18 cm piece of silicone rubber tubing (i.e. i.d. 0.63 mm, o.d. 1.19 mm) is expanded by placing it in xylene in a fume cupboard for 20 min. The xylene is washed out with sterile water. The expanded

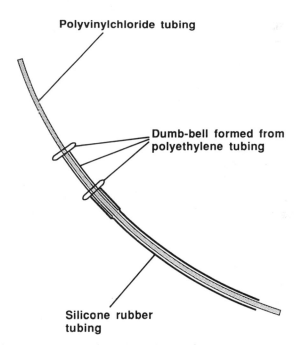

Polyvinylchloride tubing

Dumb-bell formed from polyethylene tubing

Silicone rubber tubing

Fig. 133. "Mixed" catheter used for catheterising the femoral artery.

tubing is passed over a 25 cm length of medical grade polyvinylchloride
tubing (PVC, e.g. Tygon) i.d. 0.38 mm, o.d. 0.76 mm (see Equipment
Index) and made to lie approximately 7 cm from the end of the PVC
tubing. The silicone rubber tubing forms a protective sheath and helps
prevent kinking of the PVC tubing. A dumb-bell is constructed from
2 × 1.0 cm lengths of polyethylene tubing (i.e. i.d. 0.86 mm, o.d.
1.27 mm) fused together using heat from a soldering iron. This fusing
process is performed in the presence of a steel rod which passes through
the tubing, and produces a raised edge where the two pieces of tubing
are joined. One end of the joined tubing is then flared on the soldering
iron to complete the formation of the dumb-bell. The dumb-bell is cut
to approximately 1.5 cm. The dumb-bell is passed over the PVC tubing
and the unflared end is pushed under the silicone rubber tubing at its
junction with the PVC tubing while the rubber tubing is still expanded.
The dumb-bell becomes anchored when the expanded rubber tubing
contracts.

To catheterise the femoral artery, it is exposed high up in the region
of the groin (Fig. 134) and separated from the femoral vein and femoral
nerve by blunt dissection for several millimetres. A loose ligature (first
half-hitch) is placed anteriorly and the artery is occluded by a ligature

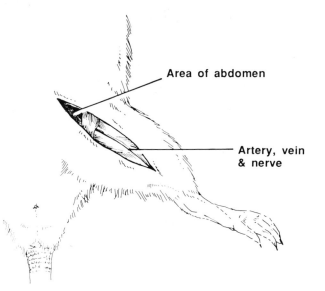

Area of abdomen

Artery, vein
& nerve

Fig. 134. Site of incision for femoral artery catheterisation.

placed posteriorly. A small clamp (e.g. Scoville Lewis clip) is placed over the artery anterior to the ligature (Fig. 135). A small incision is made in the artery and the PVC ("mixed") catheter, filled with heparin–saline (20 units heparin/ml) and attached to a heparin–saline-

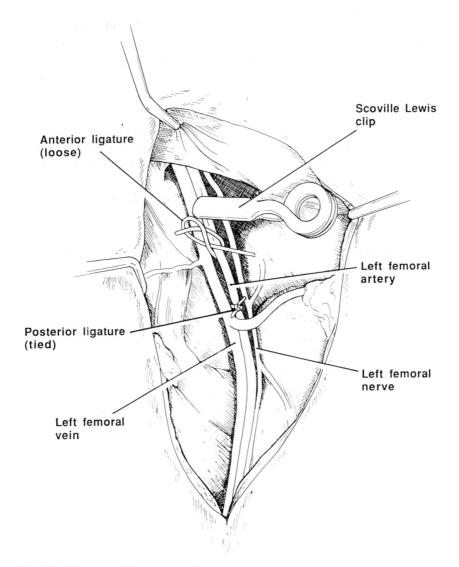

Fig. 135. Placement of ligatures and clamp.

filled syringe, is inserted. Because the femoral artery is smaller than the diameter of the PVC tubing in some animals, it may be necessary to reduce the tubing diameter by manually stretching that part of the catheter to be inserted. The anterior ligature is then loosely tied and the clamp is removed while simultaneously advancing the catheter into the dorsal aorta. The tip of the catheter, which should be square (not bevelled), should lie a reasonable distance within the aorta but below the renal veins. Once in place the anterior ligature is tied and the free ends of the posterior ligature are also used to secure the catheter. Any blood which appears in the catheter is flushed back. Implanted lengths of tubing approximately 3.5–4.5 cm are adequate for rats in the weight range 200–400 g. The catheter is gently curved and further secured by a ligature placed round the dumb-bell and surrounding tissue (Fig. 136). The catheter is passed via a long trocar to the back of the leg and thence subcutaneously up to the dorsal nape of the neck where it is secured to the skin, a pin placed in the end to maintain the "heparin lock" and then exteriorised. In practice, it may be found more convenient to feed the catheter from the neck to the incision site in the leg so that it is in place before catheterisation is carried out.

In some strains of rat (e.g. hypertensive rats), an ambulatory problem may occur because of cutting off the arterial blood flow to the leg. Generally there is enough collateral circulation to circumvent this. However, if difficulties do arise, subsequent operations should be carried out by first ligating to occlude the femoral vein before catheterising the artery. It has been suggested that this forces open collateral vessels and decreases the number of postoperative problems in the leg.

For both direct and indirect methods of catheterisation the rats are caged singly. Use of the catheter is made either while the rat is manually restrained or restrained by a suitable apparatus (see 17). The usual procedures will have to be taken to maintain catheter patency (see 22.2).

22.6. Vascular Access Ports

The main disadvantages of exteriorised vascular cannulae are that they have a known liability to becoming infected and they have usually to be protected in some way to prevent the animal damaging or removing them. The vascular access port (VAP) (Fig. 137) is a subcutaneously implanted device which allows long-term vascular access without the

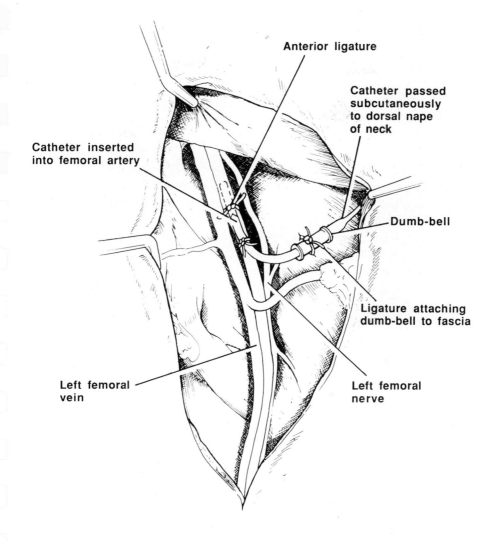

Fig. 136. Placement of mixed catheter into femoral artery—note position of dumb-bell.

above risks or with these greatly reduced. It is not a substitute for the cannula exteriorisation method in every case but it may be found useful in selected instances. The VAP is obtained commercially (see Equipment Index). It consists of a plastic dome-shaped reservoir body on an

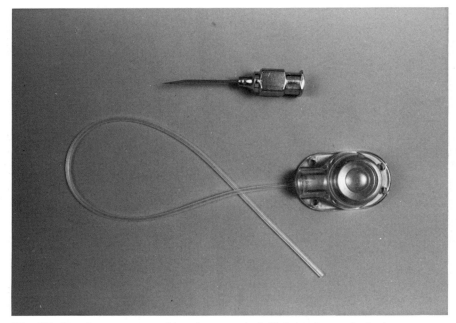

Fig. 137. Vascular access port with catheter attached. Huber point needle (top).

elliptical plastic flange base with a central compressed self-sealing silicone rubber diaphragm on top of the reservoir. The flange is pierced by several holes which are used for attaching it to underlying fascia. Connected to the port is a silicone rubber or Hydrocath™ (see 22.1.1) cannula which can be obtained complete with retention cuffs if desired. A range of cannula sizes for use in the rat can be obtained, from an internal diameter of 0.3 mm to an external diameter of 1.7 mm. The central diaphragm can be punctured several hundred times without leaking but it is recommended that a Huber point needle (Fig. 137) be used as its special shape prevents cutting of the diaphragm. However, with care standard bevel point needles can also be employed if desired.

For use in the rat, the VAP has an approximate dimension of 2.5 cm × 1.4 cm with a reservoir volume of 0.12 ml. It can be sterilised by autoclaving, by exposure to ethylene oxide gas or by disinfectants. For use, the sterile VAP is placed within a subcutaneous pocket prepared between the shoulder blades when vessels in the neck are to be cannulated, or in the lower lumbar region or flank for cannulation of the femoral vessels. It is important that the VAP lies to one side of the suture line and not under it. The flange is sutured to the underlying muscle. The cannula is tunnelled under the skin using a wide-bore needle or trocar or forceps (see 22.2) to the vascular site. The reservoir body and cannula are filled with a heparin–saline solution (20 units heparin/ml) and the blood vessel is cannulated (see above). It is advisable to leave a stress loop if retention cuffs are not used to secure the cannula. The vessel and VAP incision sites are closed. The VAP system in place should be flushed through with saline and a heparin lock established (see 22.2). When the animal is to be used it is either restrained manually or in a suitable restraint apparatus. The hair over the VAP is removed and the dome of the reservoir diaphragm is palpated with the fingers using sterile gloves. The skin is swabbed with antiseptic and a needle attached to a syringe or a three-way stopcock and cannula extension (in the case of blood pressure measurement—see Garner *et al.*, 1988) is inserted through the skin and diaphragm into the reservoir, stopping when it hits the hard reservoir base. Administration of fluids can then be made, or blood withdrawn or a suitable connection made to a transducer to record blood pressure. It will be necessary to ensure the cannula is patent and that the needle is properly positioned before this can be done (see 22.2). After use the system should again be flushed and a heparin lock put in place. When not in use the system will need to be maintained for continued patency (see 22.2). It should be noted that the VAP can be used to access not only blood vessels but any hollow structure (e.g. intestine).

23. PREPARING A CHRONIC GASTRIC FISTULA (adapted from Paré *et al.*, 1977)

The *in vivo* examination of gastric secretion and gastric function requires the preparation of a chronic gastric fistula by which gastric contents can be obtained. For best results rats should weigh 200 g or more. The components for the fistula are shown in Fig. 138. The stainless steel spool-like cannula is 12 mm long with an o.d. of 7.5 mm and has a 12 mm diameter flange top and bottom. A piece of rigid nylon or polyester mesh or a thin piece of plastic, 12 mm diameter, is attached

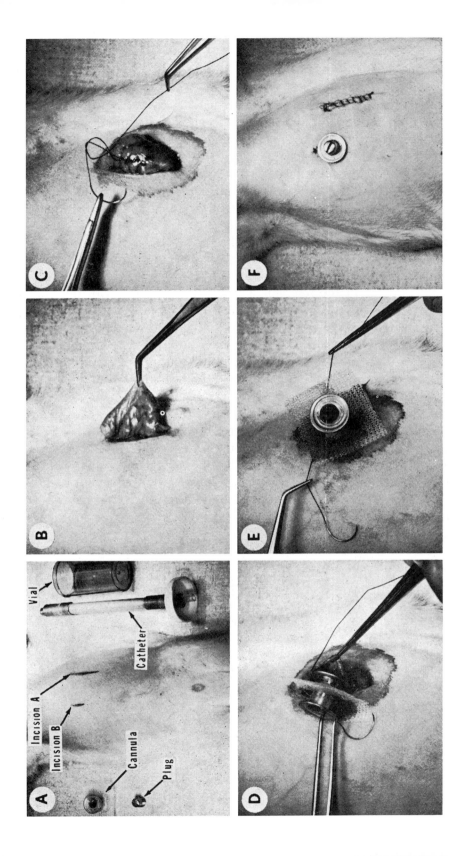

midway between the flanges to the barrel of the cannula with non-toxic adhesive or dental acrylic cement (see Equipment Index). The inner bore of the cannula is threaded at one end to take a screw-type stainless steel plug. When gastric collections are to be made, the screw is removed and replaced with a polythene or silicone rubber catheter, 140 mm long. The catheter consists of tubing, 3.3 mm i.d., which is slipped over a small hollow metal bushing with outside threads which can be screwed into the cannula. The catheter is covered with a metal tube to prevent it from being bitten through by the rat. A plastic cap is screwed onto the distal end of the tube and a 15 ml plastic vial is screwed or snapped into the cap and serves as a collection vessel for the gastric juice.

For the operation the rat is fasted for 19–24 hours, which is necessary to ensure complete evacuation of the stomach. It is anaesthetised and laid on its back with its tail towards the investigator. An incision is made in the abdomen starting 4–5 mm lateral to the midline and 4–5 mm caudal to the sternum (Fig. 138, incision A). The stomach is exposed through this incision (Fig. 138). The cannula is implanted in the anterior wall of the forestomach just above the transverse ridge. To do this, a purse-string suture (Figs 138 and 139), 2 cm in diameter, is placed into the anterior part of the stomach without passing completely through the stomach wall (i.e. placed only through the muscle and serosal layers). A small nick is made in the stomach in the centre of the purse string suture and the hole is widened by inserting and opening forceps or scissors to obtain the size required. The inner flange of the cannula is inserted into this opening (Fig. 138) so that the nylon mesh rests on the outside surface of the stomach. This mesh subsequently induces connective tissue growth and assures tightness of the cannula and fixation to the stomach. The purse string suture is drawn tight around the cannula and tied securely (Fig. 138) and the plugging screw is inserted into the cannula. A diagrammatic representation of the cannula in place is seen in Fig. 140. A small 1–1.5 cm skin and muscle incision is now made in the abdominal midline (Fig. 138, incision B). The tips of a

Fig. 138. Preparation of a chronic gastric fistula. (A) Location of lateral incision through which the stomach is pulled out and smaller incision where the cannula is secured (the components of the cannula and catheter are shown), (B) showing stomach exteriorised, (C) purse string suture, (D) inner flange of cannula being inserted into stomach, (E) cannula being secured to stomach by tying purse string suture, (F) final location of stoppered cannula in small midline incision (from Paré *et al.*, 1977).

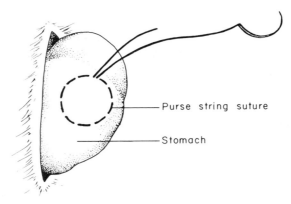

Fig. 139. Forming the purse string suture.

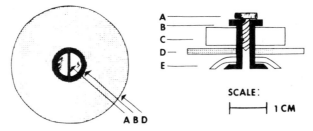

Fig. 140. A diagrammatic representation of the cannula and plastic sheet in place in the stomach: (a) screw plug, (b) cannula, (c) abdominal wall, (d) nylon plate, (e) stomach wall (from Kowalewski and Chmura, 1969).

pair of artery forceps are inserted into this incision and pushed out through the lateral incision. The outer flange of the cannula is grasped with these forceps and pulled intra-abdominally out through the midline incision where the outer flange of the cannula is externalised (Fig. 138). The lateral muscle and skin incisions are closed and one or two interrupted sutures are placed in the midline incision around the cannula to secure it in place.

After the operation rats should be housed in cages with metal grid floors. The sizes of the grid holes may have to be reduced with wire mesh if there is a tendency for the outer flange of the cannula to become snagged in them. Postoperative care requires that food is withheld for 24 hours, but the rat can be given free access to water. For the subsequent 3 days, rats are given 3 g of rat food pellets and then gradually, normal feeding conditions are restored. Animals should be allowed 18 days postoperative recovery before they are used.

To obtain gastric secretions, the food is withheld from the fistulated rat for 19–24 hours. It is then placed in a Bollman restraining cage which should be loose fitting and much larger than the usual size (see Fig. 103) to allow the rat more freedom of movement. The two centre rods of the cage floor should be made 2.0 cm apart to allow the passage of the catheter. The plug of the cannula is removed and the stomach is washed out with 4 ml of saline. The metal bush of the catheter is now screwed into the cannula and the catheter is passed downwards out of the floor of the cage and the collecting vial is attached. The gastric secretions are collected after first discarding the initial 30 min collection which will be contaminated with saline from the initial gastric lavage. Collection periods can be for at least 1 hour but if collection is to be made for more than four consecutive hours then fluid loss must be corrected by 2 ml s.c. injections of physiological saline every 2 hours. Rats with gastric fistulas can be maintained without complications for at least 1 year and probably longer.

24. HEPATECTOMY

The liver is composed of four lobes (Fig. 141). The median and left lateral lobes comprise about 70% of the liver and their removal is recognised classically as partial hepatectomy. The third lobe is the right lateral lobe which overlaps the median lobe a little on the right side. The smallest lobe is the caudate lobe and its two segments surround the abdominal part of the oesophagus. The caudate and right lateral lobes are joined across the vena cava. The lobes of the liver are bound together by folds of the peritoneum which constitute the suspensory ligaments of the lobes.

24.1. Partial Hepatectomy

The anaesthetised rat is laid on its back without restraint and with its tail towards the investigator. A midline ventral abdominal skin incision

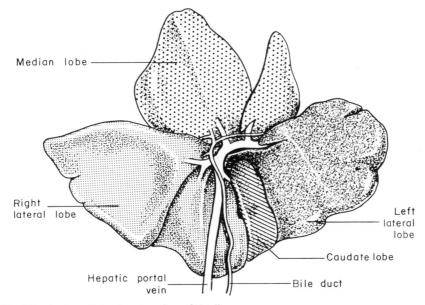

Median lobe

Right
lateral lobe

Left
lateral
lobe

Caudate lobe

Hepatic portal
vein

Bile duct

Fig. 141. A view of the dorsal surface of the liver.

is made, extending from just above the xiphoid cartilage to about half way towards the base of the tail (Fig. 142). A similar incision is made in the abdominal muscles extending from immediately below the xiphoid cartilage posteriorly to just beyond the liver. The rat is repositioned right and left (head) of the investigator and a small bolster is placed under the thorax causing the liver to fall slightly forwards away from the diaphragm. The transparent suspensory ligament attaching the convex face of the liver to the diaphragm, the falciform ligament, is cut down to the posterior vena cava with broad blunt-ended curved scissors. A piece of dry gauze is placed along the nearside edge of the skin incision and the median and left lateral lobes are moved out of the abdominal cavity onto it. Mobilisation of these lobes is neatly effected by placing both hands around the incision, as in Fig. 143, and, with the right hand, pushing the gut just posterior to the liver forwards and upwards in a concave semicircle with simultaneous light compression of the abdominal contents with both hands. This manoeuvre causes the two lobes to "pop" out and they can be grasped lightly with the hands and placed on their ventral surface on the gauze. Two other suspensory ligaments should now be revealed, one attaching the concave dorsal face of the left lateral lobe to the median blood vessels and to the stomach and one attaching the anterolateral edge of the same lobe to the dorsal peritoneum (Fig. 144). These must be cut down as near as possible to the blood vessels using the blunt-ended curved scissors. The two liver lobes are now raised vertically with the left hand and a ligature tied round them and their blood vessels at their base with a double reef knot (Fig. 145). The lobes are again laid on the gauze and several cuts made in them at their periphery to allow them to bleed on to the gauze and not into the abdominal cavity when they are finally transected (Fig. 146). The gauze is folded over the lobes which are then picked up, placed under slight tension and severed with pointed scissors above and as near to the ligature as possible. The bolster is removed and the muscle and skin incisions are closed. No special therapy is required other than a high standard of postoperative care. The liver will regenerate *in toto* within 10–20 days.

24.2. Total Hepatectomy

Removal of every vestige of the liver can only be carried out successfully (i.e. without immediate death) by a two-stage operation.

24.2.1. Stage 1

A rat of approximately 200 g is fasted for 24 hours. It is anaesthetised and laid on its back with its tail towards the investigator. A midline abdominal incision is made extending posteriorly for about half the length of the abdomen (Fig. 147). A 3 cm diameter bolster is placed under the thorax in a position which causes the liver to fall towards the diaphragm. The edges of the incision are retracted laterally and the gut is pushed over to the right of the investigator to reveal the posterior vena cava and the hepatic portal vein. The two segments of the right lateral lobe of the liver are grasped gently with the left hand and pulled vertically. This raises the vena cava slightly posteriorly and it is possible to pass curved forceps under it at a slightly diagonal angle to emerge in front and posterior to the right lateral lobe (Fig. 148). A piece of thread, size 4/0, is then grasped with the forceps and pulled back (Fig. 149) and tied round the vena cava and around a glass rod of about

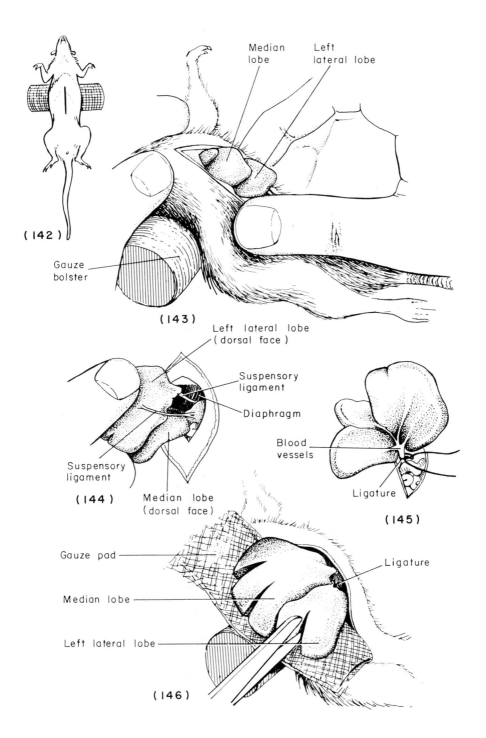

(142)

Median lobe

Left lateral lobe

Gauze bolster

(143)

Left lateral lobe (dorsal face)

Suspensory ligament

Diaphragm

Suspensory ligament

(144)

Median lobe (dorsal face)

Blood vessels

Ligature

(145)

Gauze pad

Median lobe

Left lateral lobe

Ligature

(146)

3.2 mm diameter placed on the vena cava. This ligature does not occlude the vena cava entirely because it is tied to the diameter of the rod. (For rats of 70–100 g a 1.2 mm diameter rod should be used.) Collateral blood vessels will now develop to compensate for the reduced circulation of blood in the vena cava. The loop of the duodenum is now pushed caudally to give good exposure of the hepatic portal vein. The vein is separated from the mesentery by opening a pair of forceps under it in the region of the hilum of the liver and just anterior to the inferior pancreatico-duodenal blood vessels (see Fig. 184). A piece of moistened Visking dialysis cellophane tubing, 5 mm wide and about 6 cm long, is placed round the hepatic portal vein and around a 13G needle or a piece of wire of 2.3 mm diameter which has been bent at right angles and placed along the portal vein. (The wire should be 0.79 mm o.d. for 70–100 g rats.) The cellophane tubing is tied with a reef knot and the needle or wire is removed (Fig. 150). This manoeuvre leaves the portal vein ligated to the diameter of the needle or wire and therefore causes its internal volume to be narrowed but not totally occluded. Total occlusion would kill the rat. However, total occlusion does occur slowly over the next few weeks but with the concurrent development of collateral blood vessels. The abdominal incision is now closed and the rat is allowed to recover.

24.2.2. Stage 2

After 1–2 months, when the animals have doubled their body weight and when portal and caval collaterals have developed, the rat is again fasted overnight, anaesthetised, and the abdomen reopened. If adhesions have formed between the gut and the liver, these should be carefully cleared. A bolster is placed under the rat so that the liver falls forwards. All suspensory ligaments are cut (see 24.1). Curved forceps are passed under the vena cava where it runs between the liver and the diaphragm and a piece of doubled up thread, size 0, is grasped, pulled back through and cut to form two pieces (Fig. 151). One thread is left in place to be

---

Fig. 142. Site of incision for partial hepatectomy. *Fig. 143.* "Popping" the median and left lateral lobes out of the abdominal cavity, note position of the hands. *Fig. 144.* Two of the three suspensory ligaments which have to be cut to free the liver. *Fig. 145.* Ligating the blood vessels to the median and left lateral lobes. *Fig. 146.* Cutting the lobes to allow bleeding onto the gauze.

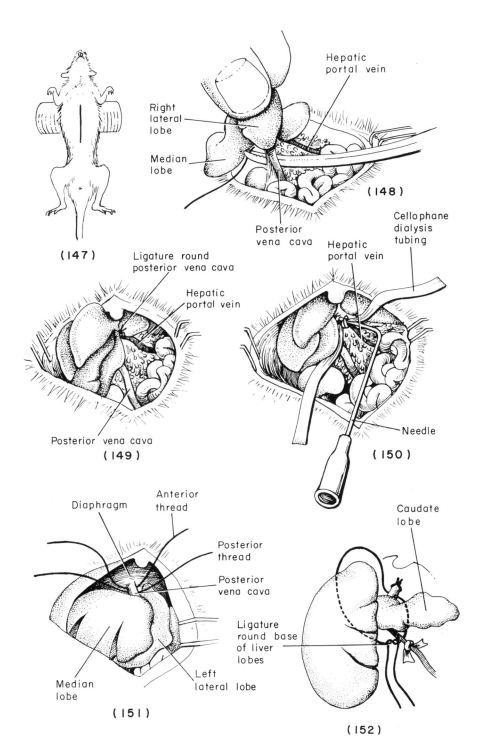

(147)

(148)
Hepatic portal vein
Right lateral lobe
Median lobe
Posterior vena cava

(149)
Ligature round posterior vena cava
Hepatic portal vein
Posterior vena cava

(150)
Hepatic portal vein
Cellophane dialysis tubing
Needle

(151)
Diaphragm
Anterior thread
Posterior thread
Posterior vena cava
Median lobe
Left lateral lobe

(152)
Caudate lobe
Ligature round base of liver lobes

tied later. In order to ligate the portal vein, the hepatic artery and any other small blood vessels and the bile duct, one end of the second, posterior thread is pulled caudally under the median, the left lateral and the caudate lobes, on the right of the investigator. This should not pose too much of a problem provided the fairly difficult process of clearing the dorsal surface of the caudate lobe of ligamentous tissue is first carried out. The other end of the thread, on the left of the investigator, is pushed posteriorly through the avascular channel under the right lateral lobe and ventral to the posterior vena cava, by lifting this lobe vertically. The complete thread should now have travelled caudally, under the liver, to end up encircling the portal vein. The two ends of the thread are tied around the portal vein just anterior to the cellophane knot (Fig. 152). A separate ligature is now tied round the vena cava just posterior to the liver and posterior to the original ligature that was placed around this vein in stage 1. Finally, the loose anterior thread round the posterior vena cava in front of the liver, is tied securely. The liver is now rendered avascular and should blanch. If not, then some small blood vessels may have been missed and must be sought and ligated. The liver is removed by cutting at the sites of the three ligatures. The muscle and skin incisions can now be closed in the normal way.

Attention to postoperative needs is necessary if the animal is to survive the maximum length of time. It should be kept warm immediately the operation is concluded and given 5% glucose in 0.9% saline hourly by s.c. injection at a dose of 600 mg glucose/kg body weight/h (12 ml/kg/h). Alternatively, the glucose can be given by continuous i.v. infusion (the tail vein is particularly useful for this, see 2.15.1), at a dose of 160 mg/kg body weight/h for the first 6 hours, then 250 mg/kg body weight/h subsequently. Most rats survive for at least 10 hours and about half will live for 24 hours. Survival beyond 36 hours is rare.

Fig. 147. Site of incision for total hepatectomy. *Fig. 148.* Placing a ligature round the posterior vena cava, a rod (not shown) must be placed on the vena cava and included in the ligature so that the vena cava is not totally occluded. *Fig. 149.* Showing the tied ligature around the vena cava. *Fig. 150.* Placing a cellophane tubing ligature round the hepatic portal vein and a needle of predetermined diameter. *Figs 151, 152.* Location of threads for ligating the posterior vena cava in front of the liver and for ligating the base of the liver lobes.

24.3. Liver Ischaemia

The liver is exposed as for partial hepatectomy and the left lateral lobe is mobilised out of the abdominal cavity and the suspensory ligaments are cut. The vascular pedicle of this lobe is ligated firmly with thread, care being taken not to injure the adjacent parts of the liver. However, it may not be possible to prevent part of the median lobe from being damaged because of some common vascularisation between the two lobes. The ligated left lateral lobe is returned to the abdominal cavity and the incision closed.

The animal tolerates the operation well for about 2 days. After this time the left lateral lobe must be removed, together with any other part of the liver which may have suffered, if the rat is to recover completely.

25. HYPOPHYSECTOMY

The pituitary gland (or hypophysis) is a small pink organ found at the base of the brain. It is located in a small bony fossa, the sella turcica, which is in the sphenoid bone of the cranium. The pituitary is attached to the floor of the third ventricle by a thin stalk and is in close relationship with the optic chiasma (Fig. 153).

There are two methods for the removal of the gland. The parapharyngeal approach via a hole drilled in the base of the cranium is the classical method, but an alternative approach is to remove the gland through the ear. Both methods require special apparatus.

25.1. The Parapharyngeal Method

For this operation a foot-operated motor controlled by a rheostat, and a flexible drive shaft are required. An ordinary domestic drill can be adapted for the purpose of drilling, but a dental drill is the apparatus of choice. The flexible drive must be able to take a contra-angle handpiece which is fitted with a round bone burr of suitable size (e.g. size 9 is suitable for rats of body weight 150–200 g). Once exposed, the pituitary must be sucked out and this is done with a Pasteur pipette,

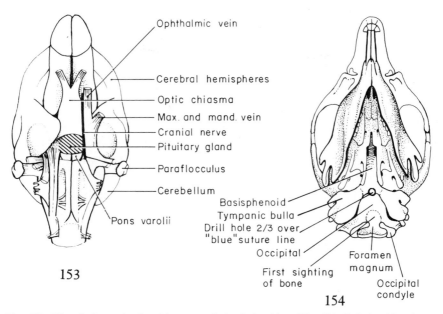

Ophthalmic vein

Cerebral hemispheres
Optic chiasma
Max. and mand. vein
Cranial nerve
Pituitary gland

Paraflocculus

Cerebellum

Pons varolii

Basisphenoid
Tympanic bulla
Drill hole 2/3 over
"blue" suture line
Occipital

First sighting
of bone

Foramen
magnum

Occipital
condyle

153

154

Fig. 153. The pituitary gland and its anatomical relationships. *Fig. 154.* Relationship of area of drill hole to floor of cranium.

curved at the end, the tip of which is of a size that is just larger than the drill hole made in the cranium to expose the pituitary. The pipette is attached to a flask with a side arm which is connected to a vacuum source. A small hole in the pipette serves as an idle air intake, and suction will only proceed when the hole is covered by a finger (Fig. 155).

For the operation the rat is laid on its back with its tail towards the investigator. The animal must be held securely on the operating board, and the head and neck must be stretched taut by means of a tooth holder. A midline incision is made along the length of the neck up to the point of the lower jaw. The trachea is exposed after division of the sternohyoid muscle (see 37.1). If the animal is anaesthetised with an inhalation anaesthetic, a bent large bore needle (e.g. 19G) is pushed into the lumen of the trachea, and anaesthesia is continued, ideally by connecting the needle to an anaesthetic apparatus via an Ayre's T-piece. Entry to the floor of the cranium is obtained by pushing a pair of curved forceps through the omohyoid muscle, on the right of the investigator, at the lower edge of the sternohyoid muscle, and at the junction formed by these muscles with the posterior belly of the digastric muscle and the fibrous arch attached to the hyoid bone (Fig. 156). Using blunt dissection, forceps are pushed downwards and towards the midline until the floor of the cranium is encountered. Muscle tissue in this area is retracted with forceps and the area widened until the white surface of the bone is seen. The usual site of the first sighting of the bone is shown in Fig. 154. The forceps are now scraped forcefully along the bone in an anterior direction for about 3 or 4 mm to remove adhering muscle tissue, until a slight slope is felt. At the top of this slope will be seen a suture line, often blue in colour. The area around and for 2–3 mm in front of this suture line is scraped free of muscle and connective tissue with forceps. Care must be taken while doing this because the pharynx lies very close in front of the suture line and must not be punctured, and blood vessels lie on either side of the suture line.

The sternohyoid muscle, the trachea and the oesophagus are retracted laterally (hence the need to intubate the trachea for maintenance of anaesthesia), and the whole area is kept widely open with as much musculature as possible down to the cranium being retracted, using strong tension (Fig. 157). All subsequent steps in the operation are now performed under a binocular microscope. Drilling is started with the bone burr held vertically and precisely in the midline and with *two-thirds* of it *in front* of the "blue" suture line (Fig. 158). During the drilling procedure drilling is stopped intermittently, and a sharp hooked needle is gently pushed into the bone in the middle of the drill hole. This is done to ascertain the thickness of the bone and how much more drilling has to be done. If this testing is not done the burr breaks through the bone unexpectedly and plummets into the brain, destroying the pituitary and producing uncontrollable haemorrhage. Any obstructive bleeding from the bone during the drilling procedure can be controlled by the use of bone wax pressed into the hole. Drilling should proceed until the needle easily breaks through the bone. In many cases, providing

---

Fig. 155. The apparatus for sucking out the pituitary. *Fig. 156.* Site of entry through the omohyoid muscle. *Fig. 157.* The "blue" suture line, note wide retraction of muscles and tissue to give a clear view. If a volatile anaesthetic is used, anaesthesia can be maintained by attaching the needle to an Ayre's T-piece. Alternatively, methoxyflurane can be placed on a gauze pad, although since gas scavenging will be difficult, this may present a safety hazard. *Fig. 158.* Hole being drilled over "blue" suture line to expose the pituitary.

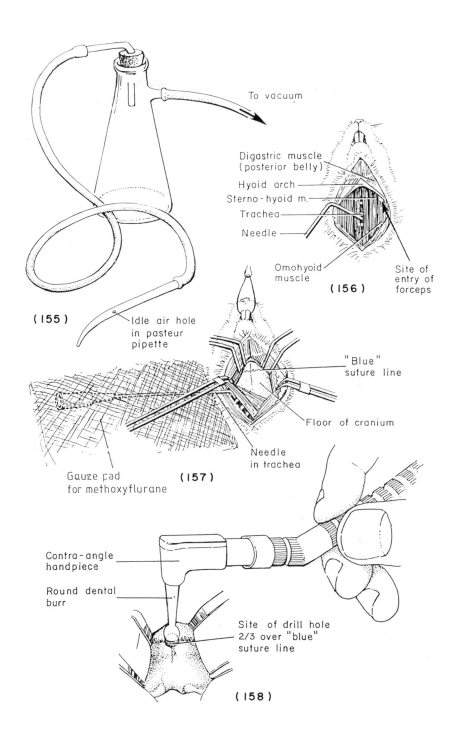

To vacuum

Digastric muscle
(posterior belly)

Hyoid arch

Sterno-hyoid m.

Trachea

Needle

Omohyoid
muscle

Site of
entry of
forceps

(156)

(155)

Idle air hole
in pasteur
pipette

"Blue"
suture line

Floor of cranium

Needle
in trachea

Gauze pad
for methoxyflurane

(157)

Contra-angle
handpiece

Round dental
burr

Site of drill hole
2/3 over "blue"
suture line

(158)

the burr is sharp, the point at which the bone is thin enough to break through the needle can be gauged while drilling since the hole starts to flatten out and circular drilling rings appear on the bottom. The drill is then removed and a circular piece of the last layer of bone is broken off and removed with the bent needle. The hole is widened in this way until most of the underlying pituitary gland is exposed. The gland is covered with a fine membrane and this is torn with the needle to free the gland. The curved tip of the suction pipette is applied to the drill hole and the gland is sucked out quickly. By sucking up some saline, the gland, often in three pieces, is wafted into the flask and can be observed, to verify removal. After removing the pituitary gland, blood may well up through the hole. This can be stopped by plugging the hole with cotton wool pledgets. The skin incision is closed after removal of the pledgets, retractors and the tracheal needle. Since a hole is left in the trachea after removal of the needle, the rat will breath some air from the s.c. pouch. In a very few cases a fatal s.c. emphysema may develop at this site. This can be avoided, if thought necessary, by sucking out any accumulated pulmonary mucus and fluid with a polythene tube at the end of the operation.

25.2. The Intra-aural Method (adapted from Gay, 1967)

The special apparatus required to perform this operation successfully, and its construction, is described in Figs 159 and 160. It can be used both for immature as well as for adult rats. For immature animals the plastic tubing covering the needle and used as a stop when the level of the needle reaches the pituitary should leave 14 mm of the tip of the needle uncovered. For rats 160–190 g body weight, 16 mm of unsheathed needle is required. In any event, the investigator should determine these measurements empirically for the strain of rat under investigation. The operation should be practised on dead animals with the brain removed but with the pituitary gland *in situ*, to determine the correct positioning of the needle over the gland.

To perform the operation, the anaesthetised animal is laid on its ventral surface with its tail towards the operator. The cartilage at the base of the external auditory meatus is cut with scissors (Fig. 161). The teeth are placed in a tooth holder and the ear bars are inserted into the ears firmly but without damaging the bones of the skull. An 18G thin-

Fig. 159. The apparatus for intra-aural hypophysectomy. The ear bars (a) are carried on a slider bar (b) 80 mm long. The point of the ear bar is 11 mm above the base board and can be rotated. The tooth plate (c) is adjustable both vertically and horizontally (adapted from Gay, 1967).

walled needle or smaller, depending on the size of the rat, mounted on a 10 ml syringe which contains 2–3 ml of water, or connected via a flask to a vacuum source (see 25.1), is inserted into the hole in the ear bars. The needle is pushed along the bony ear canal at an angle of about 5° to the horizontal in a dorsal direction produced by rotating the ear bar until it touches the medial wall of the periotic capsule where, at this point, the bone is particularly thin. In large rats the needle point is rotated against the bone to "score" it, since the bone is often brittle and would otherwise shatter. The needle is then forced through the bone with the bevel facing upwards so that the point slides under the trigeminal nerve without damaging it. The needle is pushed forwards until its travel is stopped as the plastic sheath makes contact with the ear bar. The needle is rotated so that the bevel points downwards and this should now be situated directly above the pituitary. The gland is then slowly aspirated into the water-filled syringe or into the tubing of the vacuum apparatus. The aspiration process should take about 4–7 s

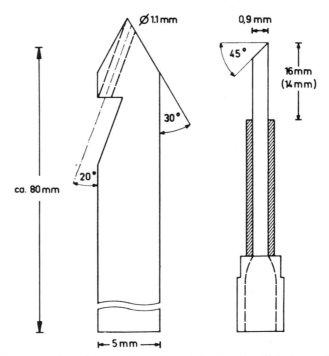

Fig. 160. The construction of the ear bar and the hole through which the needle passes; a plastic sheath (shaded) is fitted to the needle to act as a stop (from Boetzner *et al.*, 1972).

to complete. The gland should be examined to verify its complete removal. If in doubt, the entire process can be repeated through the same or the opposite ear. If desired, a needle can be inserted through both ears and the sella turcica rinsed out. If bleeding occurs this can be stopped either by pinching the ear together or by forcing a cotton wool pledget down into the ear.

The whole procedure can be carried out by an experienced person in a few minutes. Completely successful removal of the pituitary gland can be found in as high a number as 95% of the animals on which the operation is performed.

To ensure a low mortality rate from hypophysectomy carried out by either method, attention to postoperative care is important. Rats should be kept warm after the operation and given 5% glucose to drink *ad libitum* to prevent the development of a fatal hypoglycaemia. It may be advantageous to combat the stress of the operation, particularly if the

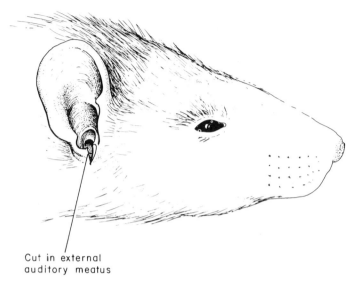

Cut in external
auditory meatus

Fig. 161. The base of the external ear is cut before insertion of the ear bars.

parapharyngeal method has been used, and to effect temporary hormone replacement, by giving an i.p. injection of cortisone (1 mg/100 g body weight) immediately after the operation, and a s.c. injection of hydrocortisone (1 mg/100 g body weight) on the second postoperative day. Hypophysectomised rats can survive in good condition for many months.

If it is desired to check the completeness of hypophysectomy, this can be done perfunctorily by gross examination of the sella turcica under a binocular microscope *post mortem*. Also, removal of the pituitary gland, particularly the anterior lobe, will result in a retardation of body growth, in thyroid inactivity, and in atrophy of the adrenal glands and the gonads. However, the use of these signs is not sufficient for testing the completeness of the operation, although they may suffice for checking on a reduced pituitary secretory activity.

The only sure way to check for the complete removal of the pituitary is by histological examination of serial sections of the sella turcica around the drill hole and that immediate part of the brain to which the pituitary was attached. It should be noted, however, that hypophysectomy by any method is not able to remove completely the pituitary stalk and pars tuberalis, and a few cells from these areas should be accepted in the histological examination.

26. HYSTERECTOMY

The rat uterus is bicornuate, each uterine horn being located on either side of the abdominal cavity and extending nearly to the lower pole of the kidneys. A single artery and vein run along the entire length of the inner side of each ovary and uterine horn and supplies blood to the ovary, uterus and cervix.

The anaesthetised rat is laid on its back with its tail towards the investigator. A midline incision is made into the abdominal cavity from just before the urethral opening to about three-quarters of the way up the abdomen (Fig. 162). Each uterine horn is pulled out and laid on

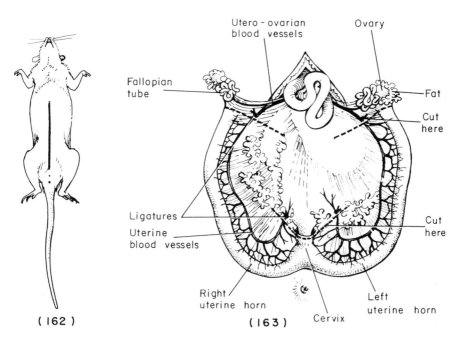

(162)   (163)

Fig. 162. Site of incision for hysterectomy. Fig. 163. Removal of the uterus and cervix.

warm saline-moistened gauze. At the anterior end the utero-ovarian vein and artery are ligated with a single ligature or staple just posterior to the collateral blood supply to the ovary so as not to interfere with the circulation to this organ (unless the ovary is to be removed as well). Posteriorly the uterine blood vessels are ligated just in front of the collateral blood supply to the cervix. The blood vessels are severed between the ligatures. Each uterine horn is cut at its junction with the Fallopian tube and the cervix is transected at its junction with the vagina (Fig. 163). Both horns attached to the cervix are removed.

Hysterectomy is often performed in pregnant rats at parturition, in order to perpetuate the rat strain as a specific pathogen-free colony. If this is the purpose of the operation it must be carried out under the strictest of aseptic conditions (see 14). Face masks and sterile gloves must be worn. The pregnant rat may be killed (rather than anaesthetised) and completely submersed in an antiseptic solution to sterilise it. As a further precaution, once the gravid uterus has been removed, it is placed immediately in a plastic box containing an antiseptic solution such as chlorhexidine/cetrimide (see 15.4.5) maintained at a temperature of 38°C. The uterus must be immersed completely. The covered box is then quickly passed into the specific pathogen-free room after first being dunked in the antiseptic solution. The foetuses are then liberated from the uterus, dried and the placenta removed a short time (but not immediately) after respiration has been initiated. The umbilicus should be ligated before being cut. It is essential at all times to maintain the correct body temperature of the offspring to ensure good survival (see "The UFAW Handbook on the Care and Management of Laboratory Animals 1987").

27. LYMPHADENECTOMY

This operation results in the removal of one or more lymph nodes. These are widely spread throughout the body, and because they are located both superficially and in deeper areas, there is no single method for their removal. The surgical approach to their removal will depend on their location and it is essential to know their anatomical distribution (see Fig. 164). Lymph node groups are classified into somatic nodes

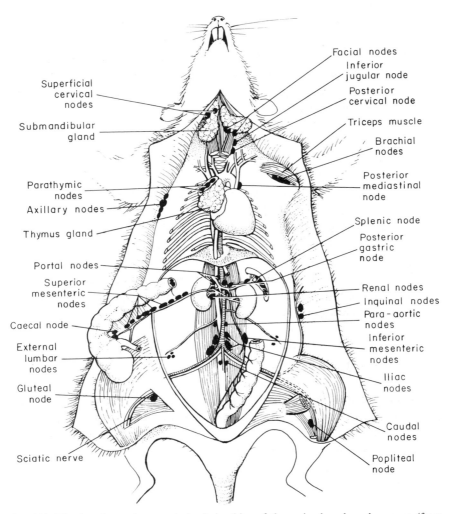

Fig. 164. The location and anatomical relationships of the major lymph node groups (from Tilney, 1971).

which drain the skin and underlying musculature (Table 11), and visceral nodes which drain primarily the thoracic, abdominal and pelvic organs (Table 12).

Lymph nodes can be visualised most easily after the i.p. injection of a 5% solution of pontamine sky blue in distilled water, 10–14 days before inspection. Macrophages coloured with the dye remain in even

minute lymphoid aggregations long after the surrounding tissues have
been cleared of the dye. Subserosal injections of hollow viscera or
subscapular injections of solid organs with vital dyes also satisfactorily
demonstrate lymphatic patterns and lymph nodes. The surgical approach
to the removal of some specific lymph nodes is now described in order
to exemplify the technique.

27.1. Superficial Cervical Lymph Nodes

These are four in number, fairly large and located on top of the large
submandibular salivary glands (two to each gland).

The rat is anaesthetised and placed on its back with its tail towards
the investigator. A midline skin incision is made in the anterior half of
the neck. The connective tissue and fat over the salivary glands and
superior cervical nodes are cleared by blunt dissection and each lymph
node is removed with forceps without effecting haemostasis with ligatures
since bleeding is inconsequential (Fig. 165).

27.2. Brachial Lymph Nodes

These nodes, usually two on either side, are found lying against the
triceps muscles near to the elbow. An incision is made from the side of
the chest to just below the elbow along the side of the forelimb. Fat
and connective tissue are cleared and the lymph nodes are plucked out
with curved forceps (Fig. 166).

27.3. Axillary Lymph Nodes

These are usually three to four in number forming a cord of tissue on
either side of the body. They are found just below the pectoralis major
muscles, lying along the chest wall near to the angle of the chest wall
with the forelimbs (i.e. the axilla). An incision is made along the side
of the chest proceeding into the axilla. Fat and connective tissue are
cleared by blunt dissection and the nodes, just under the edge of the
pectoralis major muscles, can often now be seen. Forceps are forced
through the thin cutaneous maximus muscles which overlie the lymph

Table 11. Somatic lymph nodes of the rat (from Tilney, 1971).

Region and type	Lymph node group	Location	Area drained	Efferent drainage
Head and neck peripheral nodes	Superficial cervical	Upper pole of submandibular gland	Nasolabial plexus	Central cervical nodes
	Facial	Junction of facial veins	Head and neck	Central cervical nodes
	Internal jugular	Ventral to brachial plexus	Deep cervical viscera	Central cervical nodes
Central nodes	Posterior cervical	Dorsal to brachial plexus	Peripheral cervical nodes, deep cervical viscera	Cervical duct
Upper extremity trunk, peripheral nodes	Brachial	Triceps muscle	Upper extremities chest	Axillary nodes
Central nodes	Axillary	Axilla	Upper extremities, trunk, brachial nodes	Subclavian duct

nodes and are opened to enlarge the area. The nodes can now be easily located (Fig. 167). They are removed with curved forceps after clamping them at their base with a second pair of forceps to effect haemostasis.

27.4. Popliteal Lymph Node

This single node is found in a cavity, the popliteal fossa, amidst the leg muscles at the back of the knee. The rat is placed on its ventral surface and the hind limbs are stretched out. A small longitudinal incision is made in the skin at the back of the leg behind the knee. Superficial

Table 11. Continued.

Region and type	Lymph node group	Location	Area drained	Efferent drainage
Hindquarters, lower extremity peripheral nodes	Inguinal	Flank	Thigh haunches scrotum lateral tail	Axillary nodes
	Popliteal	Popliteal space	Foot, hind leg	Lumbar and inguinal nodes
Tail	Gluteal	Sciatic foramen	Tail	Caudal, lumbar, inguinal and popliteal nodes
	Superior mesenteric	Root of mesentery	Duodenum, small bowel, caecum, ascending and transverse colon	Superior mesenteric duct to cisterna chyli
	Inferior mesenteric	Mesentery of descending colon	Descending and sigmoid colon	Inferior mesenteric duct to cisterna chyli

connective tissue is cleared, and by careful observation the area where the semitendinosus, the biceps femoris and the triceps sural muscles cross is located. This area is often denoted by a small white fatty area into which the vein running along the lower border of the triceps sural muscle disappears. Forceps are pushed into this area which delineates the popliteal fossa. The popliteal lymph node lies within the fatty tissue but is distinguished from it by being a small round discrete creamy yellow-coloured organ. The node is removed with forceps (Fig. 168).

Table 12. Visceral lymph nodes of the rat (from Tilney, 1971).

Region	Lymph node group	Location	Area drained	Efferent drainage
Thorax	Parathymic	Lateral to thymus	Peritoneal cavity, pericardium thymus	Mediastirnal duct
	Posterior mediastirnal	Paravertebral gutter	Thoracic viscera, pleural space, pericardium thymus	Mediastirnal duct
	Paravertebral	Dorsal to pulmonary vessels	Diaphragm, thoracic viscera	Posterior mediastirnal nodes
Pelvis and retroperitoneum	Caudal	Median sacral vein	Ventral tail, anus, rectum, gluteal nodes	Iliac nodes
	Iliac	Aortic bifurcation	Pelvic viscera, popliteal, gluteal and caudal nodes	Renal nodes
	Renal	Dorsal to renal veins	Kidneys, suprarenals, lumbar sympathetics	Renal duct to cisterna chyli
	External lumbar	Retroperitoneal fat pad	Fat pad, psoas muscles, pelvic viscera	Lumbar sympathetics
Abdomen	Splenic	Splenic vein	Splenic capsule and trabeculae	Posterior gastric nodes
	Posterior gastric	Gastroduodenal vein	Distal oesophagus, stomach, pancreas, splenic node	Portal nodes
	Portal	Portal vein	Liver, splenic, and posterior gastric nodes	Portal duct to cisterna chyli

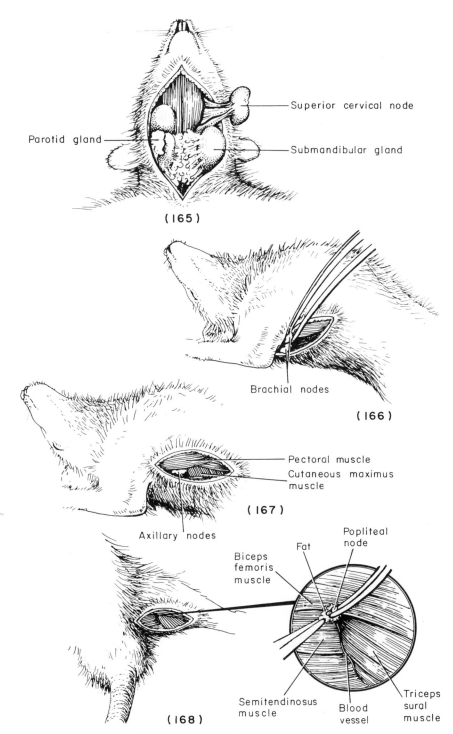

Fig. 165. Removal of the superficial cervical lymph nodes. Fig. 166. Removal of the brachial lymph nodes. Fig. 167. Removal of the axillary lymph nodes. Fig. 168. Removal of the popliteal lymph nodes.

28. LYMPH DUCT CATHETERISATION

Collection of lymph is most often carried out by catheterisation of the thoracic and mesenteric lymph ducts, and both methods will be described for the adult rat. Catheterisation of the thoracic duct is considered to be the easiest. It is possible to catheterise other lymphatic channels (see Lambert, 1965).

28.1. Thoracic Lymph Duct

The thoracic duct arises about 2 cm before the diaphragm at about the level of the left adrenal vein. It lies beside and slightly beneath the dorsal aorta on the left side, and passes anteriorly through the diaphragm into the thorax and cervical regions. Posteriorly it passes underneath the aorta as the broad cisterna chyli and gives rise to the mesenteric lymph duct (see Fig. 172). The thoracic duct is a very delicate structure and is often almost transparent and in some rat strains very difficult to see except under a dissecting microscope which, if used, greatly facilitates the performance of the operation. To make it easier to visualise and locate both the thoracic and the mesenteric ducts, 50–100 µl of a 1% (w/v) solution of a dye such as Evans blue or trypan blue can be injected into the middle members of the mesenteric group of lymph nodes which lie alongside the ilio-colonic junction (see Fig. 164). The dye colours the lymph blue immediately but only lasts for about 1 min; 5 ml of the dye can be injected intraperitoneally 30 min before the operation to achieve the same effect. Alternatively, 0.5–1.0 ml olive oil or glycerol trioleate given by gavage 1 hour before the operation will cause a white-coloured, easily identified lymph to flow in the ducts.

In addition to the basic instruments, fine forceps, a small pair of finely pointed scissors, and a long piece of polyethylene catheter of i.d. 0.58 mm and o.d. 0.96 mm will be required. The catheter must be specially shaped for use in this operation. To do this, one end is made into a U-shaped loop of about 5 mm diameter. This is most conveniently done by inserting a piece of fine wire into the end of the tubing and bending it into shape. The loop is then plunged into boiling water momentarily in order to "fix" the curve. The tubing is then cut obliquely just after it begins to curve, but in such a way as to produce only a very short bevel, which is placed to point upwards (Fig. 169).

For the operation, the rat is anaesthetised and placed on its back, right and left of the investigator. A transverse incision is made across the abdomen starting in the midline about 5 mm posterior to the xiphoid cartilage and proceeding down the left side of the rat a few millimetres posterior to the costal border (Fig. 170). The skin and muscle is retracted widely. The left kidney and adrenal gland are mobilised by gently tearing the connective tissue holding them to the dorsal abdominal wall and, together with the gut and liver, are pulled well over to the right of the animal (away from the investigator) and held in place. They should be covered by a gauze pack which is kept moistened at all times with warm saline and kept within the abdominal cavity if possible, otherwise they can be placed on top of the abdomen. It is particularly important to keep the liver well retracted and out of the way so as not to obscure the operative field (see Fig. 172). For the exteriorisation of the catheter, a stab wound is made through the dorsoposterior skin and muscle with a No. 23 scalpel blade, just caudal to the left lumbar vessel. Alternatively, a needle of sufficiently large bore to allow the passage of the catheter can be thrust through the side of the animal (see Fig. 113). The catheter is exteriorised and the curved end is left in the abdominal cavity, lying in a natural position alongside the thoracic duct (Fig. 171). The catheter should be filled with a dilute heparin–saline solution (20 units heparin/ml) and stoppered with a pin.

The thoracic duct and cisterna chyli are located. It should be noted that small vertebral arteries cross over the duct at intervals and these must not be broken. The dorsal aorta is in juxtaposition to the left psoas muscle, and they can be separated by pulling them apart with forceps or with cotton-tipped applicators. This tears the dorsal peritoneum and exposes the thoracic duct (Fig. 172). The duct is separated from the psoas muscle with fine forceps. To separate the duct from the dorsal aorta to which it is adherent by connective tissue, this connective tissue along the ventrolateral surface of the aorta is torn with forceps. A gap is then made between the aorta and the duct by opening the forceps several times with its points towards the wall of the aorta which is more resilient than that of the duct. This whole procedure is a critical one and must be done with the utmost care. Once a gap has been opened up (the position of the gap is often dictated by the positions of the vertebral arteries) a 5 mm piece of the duct is freed of connective

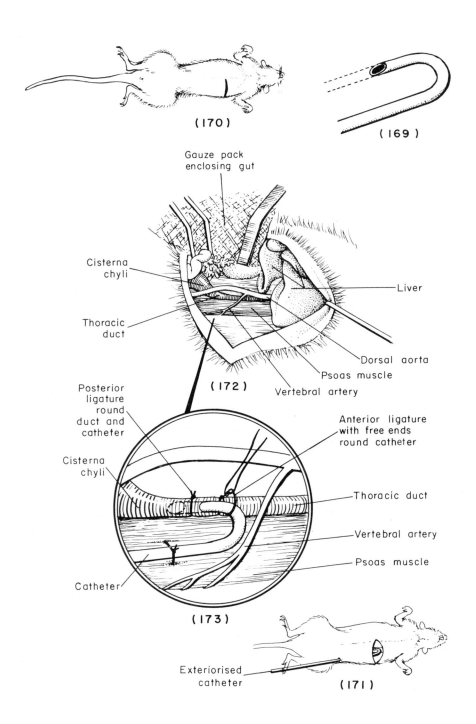

(170)

(169)

Gauze pack
enclosing gut

Cisterna
chyli

Thoracic
duct

Liver

Dorsal aorta

Psoas muscle

Vertebral artery

(172)

Posterior
ligature
round
duct and
catheter

Cisterna
chyli

Anterior ligature
with free ends
round catheter

Thoracic duct

Vertebral artery

Psoas muscle

Catheter

(173)

Exteriorised
catheter

(171)

tissue and a doubled up piece of thread pulled under it and cut to form two threads about 5 mm apart. The duct itself should be manipulated as little as possible otherwise it will constrict, but will regain its previous size if left untouched for a few minutes. The posterior thread is tied (first half-hitch) but left loose around the duct. The anterior thread is tied tightly and the duct immediately distends to about the size of the aorta. The free ends of the anterior ligature are not cut off at this point. With the points of fine scissors held vertically, the duct is partially transected (about one-third of its circumference) and should now allow the bevelled end of the catheter, with the bevel upwards, to be passed easily into the duct without force. It should be noted that once the duct has been cut it will immediately collapse and the issuing lymph, if milky or coloured, may make catheterisation difficult by obscuring the field! After the catheter is in place the posterior ligature is tied to secure it in the duct. Further anchorage of the catheter is given by tying the free ends of the anterior ligature around it and also by securing it to the psoas muscle with a single suture (Fig. 173). It is important that the catheter should be allowed to lie naturally within the duct and abdominal cavity and not made to take up a strained position by means of the various ligatures.

After the anterior ligature has been only partly tied round the catheter in the duct and before final ties are made, the investigator should ensure that the lymph is flowing in the catheter which should be unsealed at the end and placed a few centimetres below the level of the animal. If there is no lymph flow then the catheter may not have been placed in the lumen of the duct and must be withdrawn completely. It is sometimes possible to reinsert a few millimetres posterior to the first incision in the duct. After catheterisation, the internal organs are returned to their normal anatomical positions and the muscle and skin incisions are closed.

For the collection of lymph, the rat will need to be kept warm initially and restrained postoperatively in a Bollman cage (see 17.2). Where it is intended to collect lymph over several days, the harness and tether

---

Fig. 169. Preparation of the catheter for thoracic lymph duct catheterisation. *Fig. 170.* Site of incision for thoracic lymph duct catheterisation. *Fig. 171.* Exteriorisation of the catheter prior to insertion into the thoracic duct. *Fig. 172.* Exposure of the thoracic lymph duct and its anatomical relationships. *Fig. 173.* Catheterisation of the thoracic lymph duct and placement of ligatures.

system of restraint should be employed (see Section 17.3) with the catheter being exteriorised at the dorsal nape of the neck (but see below). To obtain a good lymph flow it is important to get the animal to drink as soon as possible, and to encourage it to do so the drink should be composed of 9.0 g sodium chloride, 0.5 g potassium chloride and 50 g glucose/l water. Even after the provision of good postoperative analgesia, fluid intake may be depressed, and it is therefore recommended that the animal is given a continuous i.v. infusion of Lactated Ringers solution at a rate of 1–2 ml/h. The tail vein is the most convenient to use for the perfusion, and since no protection of the i.v. catheter is required when the animal is in a Bollman cage, the tail vein can be catheterised with an over-the-needle catheter (see Section 2.15.1.2).

The flow rate of lymph that can be expected is between 0.13 and 0.7 ml/h and may at first be slow. A major problem which occurs at any time up to about 8 hours postoperatively is the almost invariable formation of lymph clots in the catheter, often at the entrance into the lymph duct. When this occurs the clots must be withdrawn as soon as possible since stoppage of lymph flow for more than 30 min is usually irremediable. The clots can be dislodged by poking a 4/0 monofilament nylon stylet (see Equipment Index) up the catheter and twisting to wrap the clot round it. The clots must not be pushed back into the thoracic duct. Unfortunately the process of removal of the clots from the catheter may not always be successful. After the first day lymph usually flows unencumbered by clot formation.

Lymph can be collected continuously for at least 1 week, and the lymphocytes in it, which are drastically reduced in number by the third or fourth day, show a viability of greater than 90% if the lymph is collected in a cooled container.

28.2. Mesenteric Lymph Duct

One hour before the start of surgery, the rat is given 0.5–1.0 ml of glycerol trioleate or olive oil by gavage. This colours the lymph white and allows the lymph duct to be seen easily. For the operation, the rat is placed on its back with its tail towards the investigator. A midline abdominal incision is made for about two-thirds of the length of the abdomen posterior to the xiphoid cartilage (Fig. 174). The intestine is brought out and placed on the right side of the abdomen and is

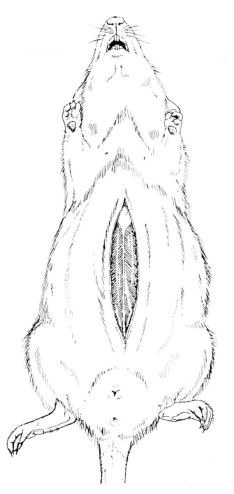

Fig. 174. Site of incision for mesenteric lymph duct catheterisation.

wrapped in gauze moistened in warm saline. The posterior vena cava and the left renal vein will now be exposed and the mesenteric lymph duct will be seen in the mesentery as a white-coloured duct about 10 mm above the origin of the left renal vein (Fig. 175). The mesenteric duct is very closely accompanied by the mesenteric artery and is attached to it by connective tissue. In most animals, the duct is present as a single vessel but sometimes it has a more diffuse appearance with interlinked channels lying on both sides of the artery. If this situation

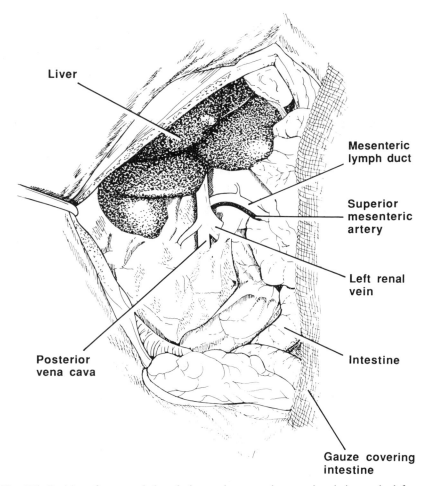

Fig. 175. Position of mesenteric lymph duct and mesenteric artery in relation to the left renal vein.

is found it is often very difficult to perform a successful catheterisation and it is advisable to use another animal. The subsequent steps of the operation are carried out with the use of a binocular dissecting microscope. The duct and artery are placed under slight tension by pressing back on the intestinal mass. Using a cotton wool pledget alternating with iris forceps, the connective tissue round the duct and artery is torn and cleared very carefully for a distance of about 10 mm, making sure that the duct is not punctured. This is a critical part of the procedure as the duct is extremely fragile. If this duct is punctured all is not necessarily lost but the escaping white lymph obscures the site and subsequent procedures are made more difficult. While clearing the connective tissue, the artery is carefully separated from the duct. Again, this may be difficult to accomplish successfully without puncturing the duct.

The next step of the procedure is to pass a trocar or similar needle which allows the passage of the catheter under the vena cava at the point of the angle the lymph duct makes with the vena cava. The trocar should lie parallel to and on top of the lymph duct. It is then passed under the vena cava and through the skin on the opposite side. If carefully performed the lymph duct is not damaged (Fig. 176). The trocar enables the catheter to be exteriorised and to lie in the correct plane relative to the duct.

A length of polyethylene tubing of approximate dimensions i.d. 0.5 mm, o.d. 1.0 mm is cut diagonally with a scalpel to produce a

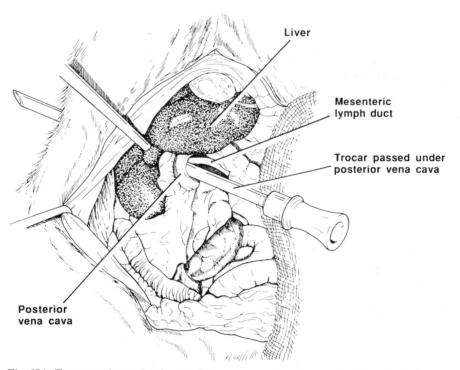

Liver

Mesenteric
lymph duct

Trocar passed under
posterior vena cava

Posterior
vena cava

Fig. 176. Trocar passing under the superior vena cava, above the mesenteric lymph duct and through the skin for exteriorising the catheter.

bevelled point. The point is blackened with indelible black marker so it can be seen within the white duct after insertion. The catheter is filled with a heparin–saline solution (20 units heparin/ml) and attached to a syringe and needle filled with heparin–saline. The catheter is passed through the trocar, which is then removed, and the tip is placed in position next to the duct. Placing the duct again under gentle traction, a small cut is made in it with fine scissors. Lymph will immediately escape obscuring the site. However, the catheter is quickly picked up with fine forceps a few millimetres behind the tip and the tip is slid along the duct until it enters at the point of section. This is done by trial and error and several attempts may need to be made. Alternatively, because the duct is so fragile it is possible to puncture it with the pointed bevelled end of the catheter. Once it enters the duct the catheter is advanced about 5 mm. A ligature is placed carefully around the duct and the catheter at a point below where the duct is sectioned. This is done using a small (e.g. 13 mm) curved atraumatic needle attached to 5/0 silk thread. The needle is passed under the duct while avoiding the artery. The first half-hitch of the ligature is tied. At this point it is important to ascertain that the catheter is correctly placed and that lymph flow is proceeding. The needle and syringe are removed from the catheter and if drops of heparin–saline are produced, lymph flow is occurring. The needle and syringe are not replaced. The second half-hitch is now tied to complete the ligature. If possible, a second ligature is placed around the top part of the duct and catheter to further secure it. Placement of this ligature will depend on the extent of the separation between the duct and the artery. A small drop of cyanoacrylate glue or tissue adhesive (see Equipment Index) is now placed on the duct at the point of entry of the catheter. This further secures the catheter and prevents leakage of lymph (Fig. 177).

The intestine is replaced in its normal position, after again making sure that the lymph is flowing, and the wound incision is closed. If lymph is being collected for a short period of a few hours, the animal can be restrained in a Bollman apparatus (see 17.2). For longer periods of collection the investigator should attempt to tunnel the catheter subcutaneously to exit at the nape of the neck and protect it by a jacket and tether system (see Section 17.3). However, there is little information as to the effectiveness of this procedure in allowing the continued collection of lymph but attempting the procedure would seem to be necessary on the grounds of increased welfare of the animal.

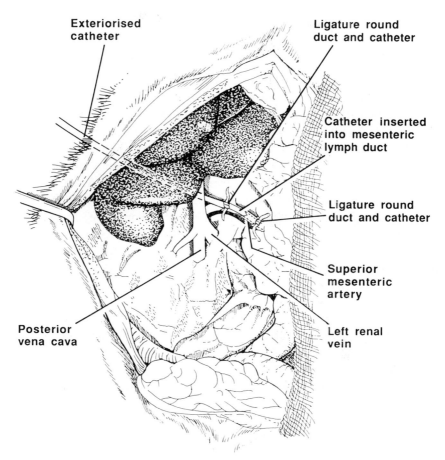

Fig. 177. Placement of the catheter into the mesenteric lymph duct.

During collection, the end of the catheter should be below the line of the animal's body to aid flow by gravity.

Clotting of the lymph in the cannula can be a frequent problem within the first few hours of collection. Some solutions to this problem and the postoperative care of the animal are described under catheterisation of the thoracic duct (see above).

29. NEPHRECTOMY

The anaesthetised rat is placed on its ventral surface, left to right of the investigator. A dorsoventral incision is made into the abdominal cavity, down the side of the rat near to the costal border of the thorax (Fig. 178). The kidney is freed of connective tissue and is pulled out gently, preferably by grasping the perirenal fat. The adrenal gland, which is attached loosely to the anterior pole of the kidney by connective tissue and fat, is gently freed by tearing the attachments. A single thread or staple ligature is placed around the renal blood vessels and the ureter as far from the kidney as possible, towards the midline, but without damaging or occluding any collateral blood vessels that may be encountered. The thread ligature is tied securely with a double reef knot (Fig. 179) and the blood vessels are transected next to the kidney which is removed. The incision is closed in the usual way.

An alternative technique is to remove the upper and lower poles of the kidney, and also much of the remaining cortical tissue, by incising with a scalpel. This results in considerable haemorrhage which is arrested by wrapping the remaining remnant of the organ in haemostatic gauze (see Equipment Index). The remaining tissue hypertrophies following the operation, and the contralateral kidney can be removed 2 weeks

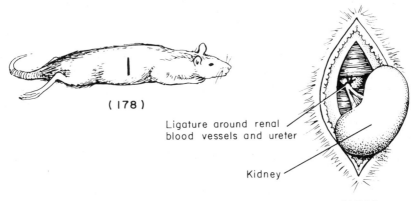

(178)

Ligature around renal
blood vessels and ureter

Kidney

(179)

Fig. 178. Site of incision for nephrectomy. *Fig. 179.* Ligation of the renal blood vessels and ureter before removing the kidney.

later. This 5/6th nephrectomy technique has been used as a model of chronic renal failure.

Unilaterally nephrectomised rats survive well with only one kidney, which subsequently undergoes slight hypertrophy. Removal of both kidneys results in the rat's death within 24–48 hours. It is possible to remove parts of the kidney, usually the upper or lower poles, by placing a ligature appropriately around it and constricting the tissue such that the portion distal to the ligature can be removed with little bleeding.

29.1. Renal Hypertension

In some scientific studies, it is necessary to produce a rat model of hypertension. This is usually carried out by occluding the renal artery almost entirely but not completely by use of special clips made either from silver tape or purchased commercially (see Equipment Index). The method has been described by Leenan and De Jong (1971) and Burns and Robbins (1972).

30. ORCHIDECTOMY (CASTRATION)

The two testes are large smooth oval-shaped organs enclosed in two muscular sacs and housed in the scrotum.

The unrestrained animal is laid on its back with its tail towards the investigator. The skin of the scrotum must be cleaned thoroughly of any faecal matter. A small median incision of about 1 cm is made through the skin at the tip of the scrotum. The s.c. connective tissue that is encountered is cleared, and the testes lying in their muscular sacs are observed (Fig. 180). These are most easily seen if pressure is placed on the lower abdomen to push the testes downwards. A small 5 mm incision is made into each sac at its tip, and the cauda epididymis is pulled out accompanied by the testis and followed by the caput epididymis, the vas deferens and the spermatic blood vessels. A single ligature or staple is placed round the blood vessels and the vas deferens, and these are severed distal to the ligature allowing removal of the testis and the epididymis (Fig. 181). The remaining piece of the vas deferens and the fat is pushed back into the sac, and each muscle incision is closed with a single suture before closing the skin incision with one or two skin clips.

If the rat is only to be *vasectomized*, the testis and the vas deferens are exposed as above, and the vas deferens is severed between two ligatures which are placed tightly around it. Each testis is returned into its sac which is closed with a suture. The two vas deferens can also be approached through a midline lower abdominal incision if desired.

31. OVARIECTOMY (OOPHORECTOMY, SPAYING)

The two ovaries are small, round, irregular-shaped organs found on each side of the abdomen, a little below the kidneys. Each ovary is

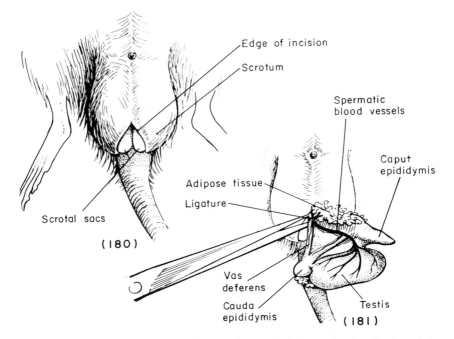

Fig. 180. Incision into the scrotum. *Fig. 181.* Removal of the testis after ligation of the spermatic blood vessels and vas deferens.

attached to one uterine horn of the bicornuate uterus via a highly convoluted and tightly packed Fallopian tube (oviduct). The approach to the removal of the ovaries is similar to that for removal of the adrenal glands (see 19).

The anaesthetised animal is laid on its ventral surface with its tail towards the investigator, a small midline dorsal skin incision is made approximately half way between the middle of the back (the hump) and the base of the tail (Fig. 182, i.e. a little more posteriorly than that made for adrenalectomy). Entrance to the peritoneal cavity is gained in the same way as for adrenalectomy. However, the muscle incisions are made half to two-thirds of the way down the side of the body. If the muscle incisions have been correctly placed, the ovaries, surrounded by a variable amount of fat, will be found underneath or within easy reach. In old rats the ovary is often totally obscured by fat, but this usually acts as a landmark and should present no problems since this fat is freely moveable. It should not be confused with the fat which will be found under and attached to the midline spinal muscles and which is not freely moveable. This latter is encountered if the muscle incision is made too dorsally and could be confusing if not recognised.

The ovary should be pulled out through the muscle incision by grasping the periovarian fat. The ovary itself must not be touched otherwise small pieces may become detached and will reimplant and carry on normal function. With pointed scissors the junction between

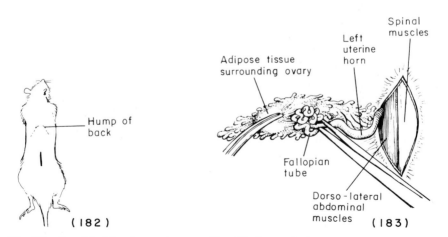

Fig. 182. Site of incision for ovariectomy. Fig. 183. Removal of ovary and Fallopian tube.

the Fallopian tube and the uterine horn, together with all accompanying blood vessels and fat, is severed with a single cut (Fig. 183) and the horn returned into the abdominal cavity. Where there is considerable fat obscuring the ovary it is safer first to cut separately the connection between the Fallopian tube and the uterine horn, and then to cut through the ovarian blood vessels and accompanying fat. In the rat, irrespective of age, there is rarely any necessity to observe haemostasis or ligate the blood vessels during the operation. Bleeding is usually slight and inconsequential and soon stops of its own accord. The muscle incision requires no suturing unless it has been inordinately large when a single suture will suffice.

If the rat is required only to be sterilised this can be achieved simply by severing the junction between the Fallopian tube and the uterine horn.

32. PANCREATECTOMY (adapted from Gay, 1965)

The pancreas in the rat is a large diffuse organ situated in the mesentery, and consequently its removal is difficult. It can be divided anatomically into three parts, which makes it convenient to describe its complete removal in three stages. A diagrammatic view is presented in Fig. 184. Particular note should be made of the vascular system and of the anterior and posterior large pancreatic ducts which carry the exocrine secretion respectively from the gastrosplenic and duodenal portions of the pancreas to the common bile duct. The biliary portion is composed of many separate lobules located along the length of the bile duct, each of which has its own duct draining into the bile duct.

Removal of the gastrosplenic and duodenal portions removes about 95% of the pancreas and is considered to be partial pancreatectomy. Removal of the biliary portion as well removes 99.5% of the organ, but the remaining 0.5% is irremovable. However, this is now considered to be total pancreatectomy and the operation for total removal will be described here.

The use of a low power dissecting microscope during much of the operation is recommended. In addition to the normal instruments, some of which should be of a fine calibre, a good supply of cotton-tipped

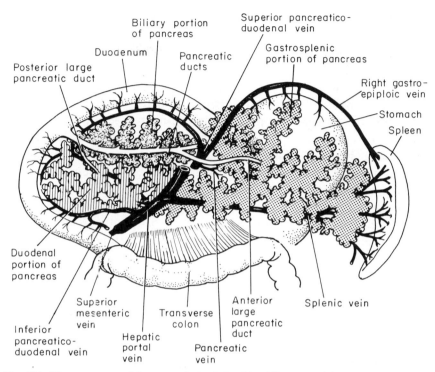

Fig. 184. The pancreas and its anatomical relationships (diagrammatic).

applicators is required. Rats should be fasted for 24 hours before the operation.

Access to the pancreas is gained through a midline abdominal incision, the incision extending about two-thirds the length of the abdomen posterior to the xiphoid cartilage (Fig. 185). The stomach and spleen are pulled out and placed onto moistened gauze on the chest wall. The duodenum is pulled out and to the left of the investigator, and the transverse colon is mobilised gently backwards after carefully cutting through the mesentery which restricts its movement. The hepatic portal vein and the root of the splenic vein coming off it should be visible (see Fig. 184). The rat is now placed with its head pointing to 7 o'clock (19:00 hours, Fig. 186). Keeping the fingers and the abdominal contents moist with saline, the spleen is held with the thumb and index finger of the left hand, with the index placed under the full length of the splenic blood vessels. With the cotton-tipped applicators and a rubbing action, the pancreas is pushed off the splenic vessels while these lie on the index finger, using a sideways movement towards the head and working up from the root of the splenic vessels towards the spleen. This removes all pancreatic tissue from the blood vessels supplying the spleen. This pancreatic tissue (the gastrosplenic portion) should now be in one mass held by the gastrosplenic ligament which is a very fine membrane attached to the spleen. The ligament is separated from the spleen and it, together with the cord of pancreatic tissue, is lifted onto the stomach (Fig. 187). The spleen is returned into the abdominal cavity after any pieces of pancreatic tissue which have become separated from the main mass are picked off with forceps.

The rat is rotated so that its head is pointing to 12 o'clock (12:00 hours, Fig. 188). The gastrosplenic cord of pancreatic tissue is taken between the left thumb and index finger and, starting at the cardiac end of the stomach, is separated from the stomach and the gastroepiploic blood vessels with forceps and cotton-tipped applicators (see also Fig. 184). When the end of the pyloric part of the stomach is reached, the band of tissue is placed on the index finger and is rubbed away from the abdomen to separate pancreatic tissue from the enclosed anterior large pancreatic duct. After exposing the duct it is cut. The band of tissue is removed and this completes stage 1 of the operation.

For stage 2, the rat is turned so that its head is pointing to 2 o'clock (14:00 hours, Fig. 189). The transverse colon should be pulled back to give good exposure of the hepatic portal vein. The duodenum is pulled to the left of the investigator and is held in the left hand with the index finger under the whole length of the bile duct. With cotton-tipped applicators and starting at the duodenal end of the bile duct and working towards the liver, the pancreas is pushed sideways off the bile duct down towards the inside of the abdominal cavity. About 2 mm from the duodenum the posterior large pancreatic duct will be encountered, and must be cut near to the bile duct after ligation with fine thread. As the little lobules of the pancreas next to the bile duct are encountered, their ducts are broken by pulling them apart with two opposing pairs of forceps. About half way between the duodenum and the liver, the cut end of the anterior large pancreatic duct will be found. This should now be ligated and cut again close to the bile duct, after thoroughly removing pancreatic tissue surrounding it and the superior pancreatico-duodenal vessels.

The animal is rotated so that its head points to 8 o'clock (20:00 hours,

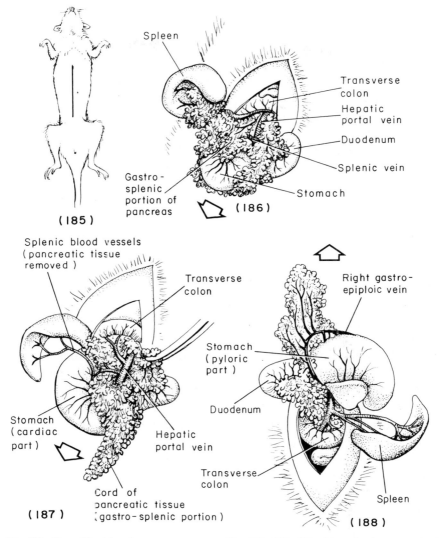

Fig. 185. Site of incision for pancreatectomy. *Figs 186, 187, 188.* Stages in the removal of the gastrosplenic portion of the pancreas.

Fig. 190). The duodenum is again taken in the left hand and pulled to the left of the investigator. The index finger is placed under the bile duct. The remaining pancreatic tissue between the curve of the duodenum and the bile duct (i.e. the biliary portion) and along the bile duct itself is now removed. Near the liver, the pancreas is difficult to see if the bile duct is not pulled out a little. At this stage of the operation the gastrosplenic portion of the pancreas and the pancreatic tissue in the biliary and most of the duodenal portions will have been removed.

For the final stage of the operation, the rat is positioned so that its head is pointing towards 9 o'clock (21:00 hours, Fig. 191). The duodenum is held between the left thumb and index finger, and the latter is placed under the inferior pancreaticoduodenal vessels. The peripheral edge of the pancreatic tissue along the back, dorsal part of the abdomen is first lifted with fine forceps and then with cotton-tipped applicators. This pancreatic tissue and that along the inferior pancreaticoduodenal vessels are pushed up towards the duodenum. After cleaning half way to the duodenum, the other half of these vessels are cleaned working from the opposite direction. The pancreatic tissue freed from these vessels is lifted posteriorly. Starting at the junction of these vessels with the duodenum, the pancreatic tissue lying along the duodenum is pushed towards the mesentery and then, when freed, the removal of the duodenal portion of the pancreas is completed. The abdominal cavity is carefully inspected for pancreatic remnants before being closed.

For *partial pancreatectomy*, the investigator should remove only the gastrosplenic and duodenal portions of the pancreas, modifying appropriately the above description of the operation for total removal.

During the operation care should be taken not to break the larger blood vessels, particularly the superior mesenteric vein issuing from the hepatic portal vein, the latter vein itself and the splenic, the gastroepiploic and the pancreaticoduodenal veins. It is less critical if some of the smaller pancreatic veins are torn, and this is probably inevitable. It is important that the vascularisation of the stomach and duodenum is not impaired to any extent.

Specialised postoperative care is necessary if the animal is to survive. Totally pancreatectomised rats live only 1–2 days if untreated, dying in diabetic coma. It is therefore necessary to provide insulin, and the course of injections recommended is shown in Table 13. Because of the need to give insulin, food intake has to be controlled at all times postoperatively, and this is achieved by tube feeding a liquid diet (see 56). It will be necessary to insert surgically a polyethylene or silicone rubber feeding tube into the oesophagus, and exteriorise it at the dorsal part of the neck where it is stoppered. The exposed part of the tube is

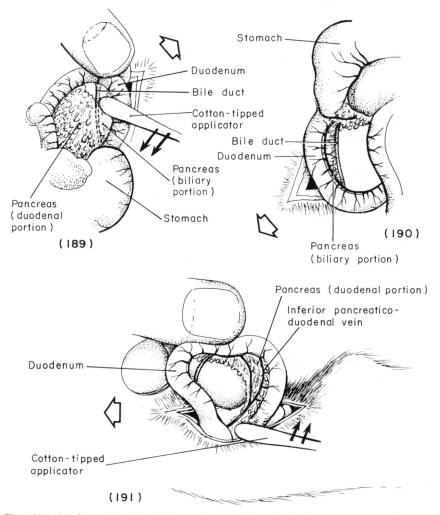

Figs 189, 190. Removal of the biliary and part of the duodenal portions of the pancreas.
Fig. 191. Removal of the final part of the duodenal portion of the pancreas.

connected to a syringe filled with the liquid diet when feeding is to be carried out. Since totally pancreatectomised animals show some degree of impairment of food absorption, the administration of pancreatin with the liquid diet is required to overcome this. An appropriate schedule is shown in Table 13. Partially pancreatectomised rats require less insulin (about 12 units/day) and no pancreatin since food absorption is normal.

Table 13. Schedule of postoperative treatment for totally pancreatectomised rats.

Postoperative day	Regular insulin (units)[*]	Food–pancreatin suspension (ml)[†]
0 p.m.	0	0
1 a.m.	0.5	0
p.m.	2.4	2[‡]
2 a.m.	4	3
p.m.	5	4
3 a.m.	6	5
p.m.	9	7
Then daily a.m.	9	7
p.m.	9	7

[*]Insulin is injected 1–2 hours before the morning meal and 15–30 min before the afternoon meal. The afternoon meal is given 6 or more hours after the morning meal.
[†]This suspension contains 0.5 g of diet and 5 mg of pancreatin in each millilitre. The pancreatin is added to the food just before feeding.
[‡]Drinking water is withheld from the rats postoperatively until this meal.

It is necessary that the blood glucose levels of pancreatectomised rats are monitored in order to establish a normal pattern for the rat under study. As a precaution against infection, benzylpenicillin (see 58) is injected for the first 2 days postoperatively.

33. PINEALECTOMY (adapted from Hsieh and Ota, 1969)

The pineal gland is a tiny (1–2 mm) round, rather translucent organ, situated on top of the brain at the confluence of the superior sagittal and transverse sinuses, between the cerebral hemispheres and the cerebellum (Fig. 192).

The rat is anaesthetised and laid on its ventral surface with its tail towards the investigator. The head is held firmly in a special apparatus which is described in Fig. 193. A midline incision is made in the scalp skin between the ears and the eyes, and the edges are retracted after clearing the underlying connective tissue. The transverse and superior sagittal sinuses can now be seen through the bone. Using a dental drill with a contra-angle handpiece and a No. 1/2 fissure burr (Fig. 194), a fine groove is cut into the bone along line "a" for about 4 mm and

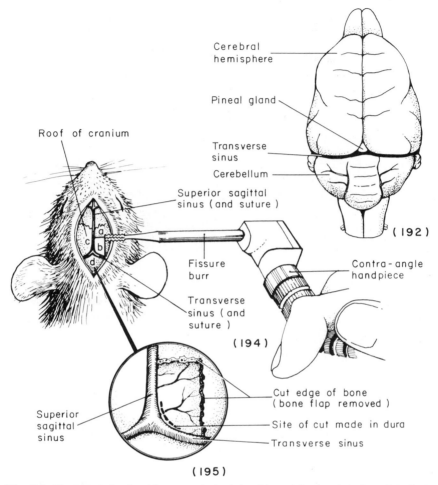

Fig. 192. The pineal gland and its anatomical relationships on the dorsal surface of the brain. *Fig. 194.* Cutting a flap in the dorsal part of the cranium with a fissure burr. *Fig. 195.* Site of cut into dura before removing pineal gland (adapted from Hsieh and Ota, 1969).

along line "b" for about 6 mm, until the bone is just broken through. This produces a flap of bone. Sides "c" and "d" cannot be cut because they overlie the blood sinuses, so they are heavily scored using either the burr or another sharp instrument. The flap is now raised with forceps at the junction between sides "a" and "b", broken off along sides "c" and "d", and carefully detached from the underlying dura

Fig. 193. An apparatus for holding the head steady during pinealectomy. The lower plate (a) is 90 mm long and the end piece (a_1) is 37 mm long and 10 mm wide, the two raised flanges (a_2) at the tip are 2 mm high, the plate is placed in the mouth along the palate and the incisors fit into the hole a_1. The horizontally adjustable top plated (b) is 45 mm long, and 20 mm wide at the bottom part of the curved end piece (b_1). This end piece fits over the rat's snout and is screwed down to hold the head steady (by kind permission of Dr D. Carter).

(Fig. 195). The dura is cut with a sharp needle in an arc-shaped line (dotted line in Fig. 196) in the area of the confluence of the superior sagittal and transverse sinuses. The dura is raised with iris forceps and a second pair of iris forceps approaches the pineal gland at the confluence of the sinuses and removes it. Any bleeding is controlled with cotton wool pledgets. The flap of bone is replaced after dusting the dura with an antibiotic powder (see 58), and the skin incision is closed with a continuous stitch.

Confirmation of the completeness of the removal of the gland is obtained by microscopic inspection of the area *post mortem*. The operation is in no way detrimental to normal survival.

34. RENAL TRANSPLANTATION

This technique has been included as an example of organ transplantation in the rat, as well as to illustrate the general principles of microsurgical technique (see Green and Simpkin, 1988). Organ transplantation depends on the successful anastomosis of the relevant blood vessels, the diameter of which may be less than 1 mm. To cope with such small vessels, very fine instruments are required, such as those used in paediatric, ophthalmic and neurosurgery. Those needed for renal transplantation are the instruments shown in Figs 79, 81, and 82. In addition a curved haemostat may be required, such as a pair of Wright's orbital forceps or a De Bakey atraugrip occlusion clamp for paediatric surgery, both of which require some modification to produce an adequate smooth-surfaced, spoon-shaped curve to clamp the dorsal aorta and posterior vena cava while leaving a sufficient area of the vessel walls available for the anastomosis. Microsurgical work is facilitated greatly by the use of a binocular dissecting microscope or a pair of purpose-built spectacles incorporating magnifying lenses with a 2.25 diameter magnification and a 25 cm focal length.

Renal transplantation can be carried out either by an end-to-end anastomosis of the renal blood vessels, or by anastomosis of the ends of the donor renal blood vessels with the sides of the dorsal aorta and posterior vena cava of the recipient. The method of choice will depend on the preference of the investigator and whether the donor animal is to remain alive. Both methods will be described.

34.1. End-to-end Anastomosis

34.1.1. Preparation of the Donor Kidney

The anaesthetised rat is placed on its back with its tail towards the investigator. A long midline incision is made into the abdominal cavity. The liver is retracted against the diaphragm and the gut is wrapped in a warm saline-moistened gauze pack and displaced to the left of the investigator, if the left kidney is to be prepared, outside of the abdominal cavity. The renal artery and vein are found, and any attached collateral blood vessels (e.g. the spermatic and adrenal veins in a number of cases) are ligated and cut. By careful blunt dissection, the renal artery and vein are separated from each other along their whole length and cleaned most thoroughly of adventitia and fat. This is an important and quite difficult step during which other small attached blood vessels may be encountered which must be ligated and cut (Fig. 196). The walls of the renal vessels, particularly the artery, should be touched as little as possible otherwise they will constrict. Excess fat is removed from around the kidney which is freed from its connective tissue attachments.

The ureter is located and cut at its widest point, usually about 2 mm below its entry into the kidney. It must not be stripped of its adventitia or fat otherwise necrosis can subsequently occur. The stub of the ureter is now catheterised with a nylon or polythene tube, i.d. 0.28 mm, o.d. 0.61 mm, which is passed into the renal pelvis for a short distance. A small vascular clip is placed round the catheterised ureteral stub to anchor the tubing in place. The ureter must not be ligated. The tubing is cut obliquely about 2 mm below the end of the ureter to facilitate insertion into the ureter of the recipient. The animal is now ready for the removal of the kidney—but do not remove it yet. The rat can be injected with heparin to prevent clotting of the blood in the kidney, but this should be avoided if possible since it may lead to difficulties later on when clotting of the blood at the anastomotic junction is helpful to the success of the operation. The animal is now placed to one side, kept warm, and the kidney is kept moist with saline.

34.1.2. Preparation of the Recipient

The renal blood vessels are prepared as for the donor. Removal of adventitia and fat should be particularly thorough in the hilum of the kidney. It should be noted that in some rat strains the renal artery bifurcates before entering the kidney. The ureter is cut as close to the kidney as possible. The renal artery and vein are separately clamped with small vascular clips next to the dorsal aorta and posterior vena

---

Fig. 196. Preparation of the donor kidney for end-to-end anastomosis. Fig. 197. (A,B) End-to-end anastomosis of the renal artery, (C) turning the renal artery on its back to continue suturing. Fig. 198. End-to-end anastomosis of the renal vein. Fig. 199. End-to-end anastomosis of the ureter, showing the internally placed catheter.

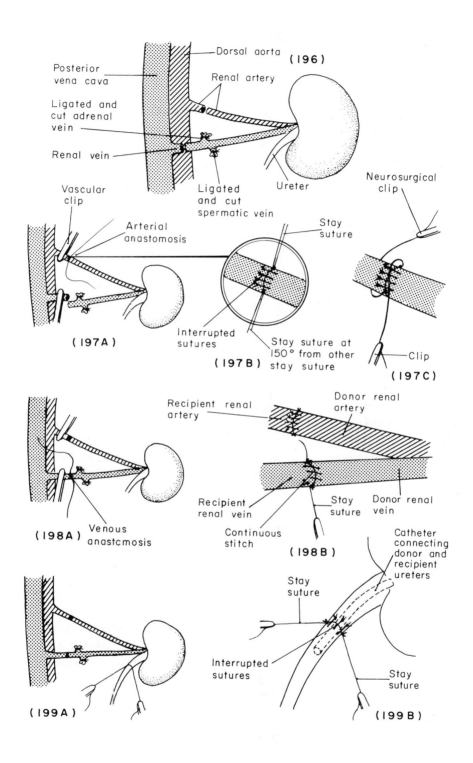

Posterior vena cava

Dorsal aorta

Renal artery

(196)

Ligated and cut adrenal vein

Renal vein

Ureter

Ligated and cut spermatic vein

Vascular clip

Arterial anastomosis

Stay suture

Neurosurgical clip

Interrupted sutures

Stay suture at 150° from other stay suture

Clip

(197A)

(197B)

(197C)

Recipient renal artery

Donor renal artery

Recipient renal vein

Stay suture

Donor renal vein

Venous anastomosis

Continuous stitch

(198A)

(198B)

Catheter connecting donor and recipient ureters

Stay suture

Interrupted sutures

Stay suture

(199A)

(199B)

cava respectively (see Fig. 197). The two renal vessels are cut cleanly across near to the kidney. If the artery bifurcates, then either one branch can be ligated and cut while the other is cut without ligation or, if the bifurcation occurs very near to the kidney, the artery can be severed just before the bifurcation. The cut ends of the artery and vein are irrigated with saline to remove any blood. The artery should not be held between the tips of the forceps, but only manipulated with the side of the forceps, to reduce the possibility of its constricting. However, if it does constrict, it will be necessary to dilate its tip by gently inserting the ends of the forceps and opening them fractionally. The kidney is discarded and the recipient is now ready to receive the donor kidney.

The renal artery and vein of the donor are clamped near to the dorsal aorta and posterior vena cava with artery forceps, and cut cleanly across just below the forceps. The cut ends are irrigated with saline, and the kidney is placed in the position vacated by the recipient kidney. Then the ends of the renal arteries and veins, and the ureters of the donor and recipient rats, are brought together in preparation for suturing.

34.1.3. Suturing
34.1.3.1. The renal artery. Using 10/0 monofilament nylon suture material attached to a 6 mm curved atraumatic needle, or the newer 14 micron monofilament nylon suture bonded to a 3 mm needle, two stay sutures are placed in the opposed ends of the arteries at about 150° to each other, leaving a long tail of thread to each suture. The artery can be held lightly while suturing. The placing and tensioning of these two stay sutures causes the opposed arterial surfaces to gape slightly which facilitates suturing. The tails of the two threads are clamped with small serrated neurosurgical clips (e.g. Scoville Lewis clips, see Fig. 126) and positioned on the fat and other tissue in the region of the kidney so that when the threads are tensioned the renal artery is slightly raised. As their name implies, stay sutures are for anchoring and positioning the vessels during suturing. The top of the opposed arteries are now sutured with four to ten interrupted stitches depending on the size of the suture material (Fig. 197). The suture material must not pass right through the walls of the arteries where it would act as a focus for the formation of clots in the lumen, but must only pass into and out of the outer layer (intima) of the arteries.

When suturing, the stitches must not be pulled too tight but only just sufficiently to oppose the vessels' edges. Moreover, the thread must not be pulled through against the pull of the artery, but a pair of forceps should be used to prevent movement of the artery as the thread is pulled through it. This avoids the suture hole becoming too large. To suture the back of the renal artery, the right hand stay suture clamp and thread are moved *over* the artery onto the left side of the investigator while the left hand thread with the clamp removed is pushed *under* the renal artery to the right side and reclamped (Fig. 197). This manoeuvre twists the renal artery presenting the back of it. This part is now sutured with four to ten interrupted sutures. All excess thread is now removed.

At least 10 sutures of 10/0 thread (and more if 14 micron thread is used) must be placed to anastomose blood vessels of 1 mm diameter. Fewer than these would result in uncontrollable haemorrhage.

34.1.3.2. The renal vein. A similar procedure is carried out as for the artery, but only the left hand (bottom) stay suture thread is clamped initially. To suture the back of the vein the left hand thread and clip are brought over to the right side of the investigator, and the unclamped right hand thread is passed under the vein to the left side where it is now clamped. Suturing is carried out using a single continuous stitch placed round the whole circumference of the blood vessels (Fig. 198). Again, the edges of the veins must only be opposed gently, and the diameter of the vein at the anastomosis must not be reduced. It is less critical to hold the vein during suturing than it is to hold the artery.

After suturing the renal blood vessels, the clip over the renal vein of the recipient is removed. Blood will run back into the kidney and there will be a little bleeding which will soon stop. Following this, the clip on the renal artery is loosened momentarily and then replaced. There will be profuse bleeding at the anastomosis during this momentary period. The clip is again loosened momentarily and this procedure is repeated several times with a pause between each to allow the blood to clot along the line of anastomosis. If bleeding still persists after several minutes, the investigator must ascertain where the bleeding is coming from, and place another suture in the adventitia at the bleeding point. The procedure of momentary loosening of the arterial clip should continue until bleeding stops. This can take as long as 30 min, particularly if heparin was used in the donor animal. If the anastomosis has been made correctly, the kidney will become pink in colour every time the arterial clip is loosened, and also when the clip is finally removed after all bleeding at the two anastomoses has stopped. If this does not happen, then there is a fault in the revascularisation process which may be difficult to overcome.

34.1.3.3. The ureter. Bloody urine may emerge from the ureter after the renal artery clip is loosened and finally removed; this is normal. The ureter of the recipient is now catheterised with the free end of the tubing in the donor ureter. The two ureters are sutured together with about four interrupted stitches and using two stay sutures for the purposes of orientation. Size 8/0 thread is used for ureteral anastomosis. The catheter is left *in situ* and causes no interference (Fig. 199A, B). The abdomen can now be closed after replacing the gut in the correct anatomical position.

34.2. End-to-side Anastomosis

Anastomosing the blood vessels in an end-to-end fashion, as just described, is a relatively difficult procedure, but is necessary if the donor rat is to remain alive. The transplant operation is made easier if a larger and firmer area of the blood vessels is available for suturing, and this is obtained in an end-to-side anastomosis. It results in the death of the donor animal.

34.2.1. Preparation of the Donor Kidney

The preparation of the donor kidney is similar to that described in the procedure for end-to-end anastomosis, except that the dorsal aorta and posterior vena cava are also cleaned of fat and adventitia for 3–4 mm below and above the junction with the renal vessels. Care must be taken in holding the vena cava which has a thin wall. The dorsal aorta is ligated 2–3 mm below the renal artery and clamped with artery forceps 3–4 mm above the renal artery, after first ligating any collateral vessels such as the artery to the opposite kidney and the superior mesenteric artery. The dorsal aorta is now cut cleanly across, just below the clamp, and also just below the ligature. In this way a cuff of aorta is formed with the renal artery attached. Where the renal vein is attached to the vena cava, a small elliptical segment is cut out of the vena cava with curved iris scissors, after first clamping it posteriorly, so that this segment remains attached to the renal vein (Fig. 200). The ureter is prepared with a catheter as described previously (see p. 288). The kidney is removed and immediately perfused via the aortic cuff with about 10 ml of cold saline. It is then placed in a beaker of cold saline while the recipient animal is made ready.

34.2.2. Preparation of the Recipient

A right or left nephrectomy (see 29) is carried out through a midline abdominal incision, depending on which kidney is being transplanted. The dorsal aorta and posterior vena cava are cleaned of fat and adventitia for a few millimetres posterior to the renal vein. This area of the aorta and vena cava is now completely clamped with a specially prepared curved haemostat (Fig. 201; also see introduction to this operation). Blood flow in these vessels is usually completely stopped by this procedure but it is good practice to attempt to clamp the two vessels in such a way that a small flow can still be obtained. A sufficient amount of the aorta and vena cava must be available in the curve of the clamp for the subsequent anastomosis. A small elliptical opening is made with curved iris scissors in the dorsal aorta anteriorly and in the posterior vena cava posteriorly in the clamped part of these vessels. The size of the openings should match as nearly as possible the segments left on the blood vessels of the donor kidney. If scissors small enough cannot be obtained to make the elliptical openings, the aorta and vena cava can merely be slit with a sharp cutting needle, the edges of the slit being used for the anastomosis. The openings in the aorta and vena cava are irrigated with saline to remove blood and small clots.

34.2.3. Suturing

The donor kidney is placed in position and two sutures at 180° to each other using two separate curved atraumatic needles and 7/0 thread are placed in the opposed dorsal aorta and donor renal artery. After tying the knots, the free tail of each thread is clamped with small serrated neurosurgical clips and used as the stay sutures to produce slight tension on the vessel. The top of the anastomosis is sutured with a continuous stitch using the right hand needle (Fig. 202). The kidney is then flipped over to the opposite side to expose the underside of the vessels which are now sutured together using the left hand needle (Fig. 203). A similar procedure is carried out for the vena cava–renal vein anastomosis.

After suturing, the curved clamp is slowly removed while applying pressure on the two anastomoses with a moist gauze pad. The aorta is

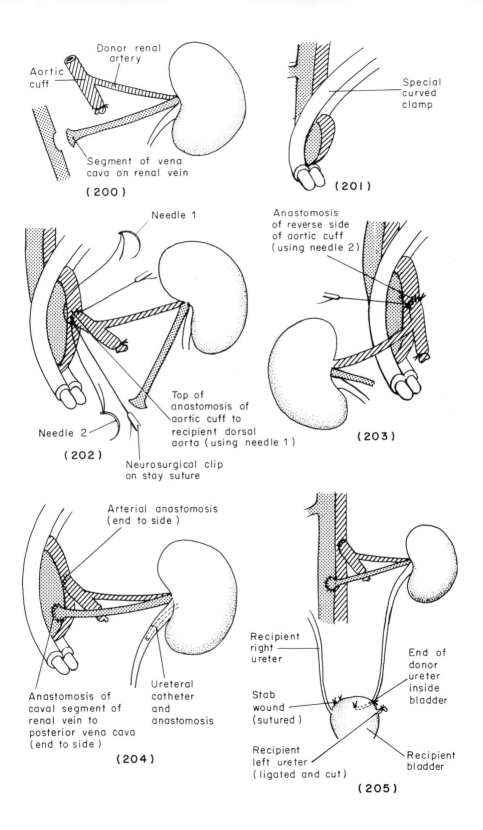

(200)

Aortic cuff

Donor renal artery

Segment of vena cava on renal vein

(201)

Special curved clamp

(202)

Needle 1

Needle 2

Top of anastomosis of aortic cuff to recipient dorsal aorta (using needle 1)

Neurosurgical clip on stay suture

(203)

Anastomosis of reverse side of aortic cuff (using needle 2)

(204)

Arterial anastomosis (end to side)

Anastomosis of caval segment of renal vein to posterior vena cava (end to side)

Ureteral catheter and anastomosis

(205)

Recipient right ureter

Stab wound (sutured)

Recipient left ureter (ligated and cut)

End of donor ureter inside bladder

Recipient bladder

immediately clamped just anterior to the anastomoses with a small vascular clip. The recipient and the donor ureters are now joined as before. The aortic clip is removed slowly while gentle pressure is again applied to the anastomoses, and this should be continued for about 5 min or until there is no further bleeding from the anastomoses (Fig. 204). Some prodding of the vessels may be necessary to get the blood flowing again. If there is clotting in the aorta, the ligature round the free end of the aortic cuff can be undone and a probe passed up the aorta and renal artery.

Some objection could be made for the use of a catheter to join the ureters and subsequently leaving it *in situ*. Although catheterisation makes the ureteric anastomosis easy and is also apparently well accepted by the rat, there is no reason why anastomosis cannot be carried out without an intervening catheter. The principle is the same as that for joining the blood vessels, as described above. However, another method can also be used, based on inserting the end of the donor ureter into the bladder of the recipient. For this procedure the donor ureter is severed close to the bladder. A small stab wound is made in the side of the recipient's bladder, and a small pair or curved forceps is inserted and pierced through the posterior wall of the bladder. The tip of the donor ureter is grasped with these forceps and is drawn into the bladder so that it fits comfortably without tension. Where the ureter passes into the bladder, a 7/0 nylon stitch is placed in the wall of the ureter securing it to the bladder wall. The short piece of ureter protruding into the bladder is anchored to the bladder wall by passing a single stitch through the bladder wall, and the wall of the ureter and this is tied on the outside of the bladder. The stab wound is closed with a short continuous stitch (Fig. 205).

The total time of ischaemia that the donor kidney undergoes must not exceed 45 min, otherwise the kidney may not function subsequently. Rats carrying transplants do not require any special postoperative care except to be kept warm for an hour or two. If it is necessary to remove

---

Fig. 200. Preparation of the donor kidney for end-to-side anastomosis. *Fig. 201.* Clamping of the recipient dorsal aorta and posterior vena cava with a curved clamp fitted with thin rubber or plastic tubing. *Fig. 202.* Suturing the top of the donor aortic cuff to the recipient dorsal aorta, note the two needles for suturing. *Fig. 203.* As for Fig. 202 but showing suturing of the underside; the kidney has been flipped over. *Fig. 204.* Complete anastomosis of the renal blood vessels and the ureter. *Fig. 205.* An alternative and direct method of connecting the donor ureter to the recipient's bladder.

the remaining kidney of the recipient, this should be done about 2 weeks later, though total nephrectomy at the time of transplant has been carried out successfully in some cases.

When transplantation is performed between members of the same inbred strain, the transplanted kidney remains healthy and the recipient shows a normal blood urea nitrogen level. When different strains are used, rejection phenomena occur by about 8 days and the kidney ceases to function. A simple way to gauge kidney function is to measure the osmolality of the urine.

35. SKIN GRAFTING

This procedure entails the placing of donor skin grafts into specially prepared graft "beds" on recipient animals. The grafts have to be kept in place with a suitable dressing so that the rat cannot gain access to the graft by biting through the dressing material. The following description is concerned only with the use of full-thickness donor skin grafts.

In addition to the basic instruments a No. 23 (straight) and a No. 12 (curved) scalpel blade with holders will be required and also a sharp pair of small curved scissors. The following accessories will also be needed:

(i) Rolls of plaster of Paris bandage about 2.5 cm wide by 16 cm long wound round small plastic spindles. These spindles can be cut from the spindle supporting wider lengths of commercially obtainable plaster of Paris.

(ii) A plasticised celloidin solution prepared to the formula: thick celloidin or pyroxylin* (10 g), camphor (2 g), castor oil (2 g), ethyl alcohol (25 ml), and ether (75 ml). The plasticiser should feel tacky between the fingers when properly prepared.

---

*Pyroxylin is also known as collodion cotton, colloxylin, xyloidin and soluble gum cotton. Alternatively, necolloidine can be used at a dry weight of 25 g in the above formula.

(iii) Tulle gras which is prepared by soaking 3.5 × 3.5 cm squares of close-woven cotton bandage in yellow petroleum jelly. Alternatively, Melolin sterile non-adherent pads can be used instead of tulle gras (see Equipment Index).

(iv) Gauze pads about 5 cm square.

(v) Covered Petri dish containing one or two circles of filter paper.

Skin grafting must be carried out using full aseptic precautions (see 14). Instruments must be boiled or autoclaved and laid on sterile cloth or paper drapes. The cotton pads, Petri dish and drapes are wrapped in aluminium foil, and the tulle gras bandage is placed in a beaker containing solid yellow petroleum jelly and covered with aluminium foil. These are now dry heat sterilised (see 15.4). The operating board is cleaned with a disinfectant.

35.1. Preparation of the Donor Skin

The donor graft is usually prepared from skin of the lateral thoracic area or from the tail. It is recommended in the literature that skin with active hair growth should not be used. However, in our experience this is not a necessary requirement.

35.1.1. Trunk Skin Graft

The anaesthetised or freshly killed rat is laid on its side, left and right of the investigator, and the fore and hind legs maintained stretched out. The hair over the lateral thorax is removed with clippers, the area is swabbed with an antiseptic solution, and the plasticiser is rubbed in using a pad of gauze and a side-to-side movement. This procedure removes any loose hair and "fixes" the rest of the hair in place (Fig. 206). The skin over the thorax is grasped with forceps and pulled upwards. A round full-thickness graft is obtained by slicing the skin with a No. 12 curved scalpel blade (Fig. 207). The graft should be roughly circular and about 2.5 cm in diameter. Some trimming can be done using scissors. This graft contains not only the skin but an attached layer of muscle and connective tissue which is known as the panniculus carnosus. The skin is then prepared by removing the panniculus carnosus. To produce such a graft, the skin is placed in a sterile Petri

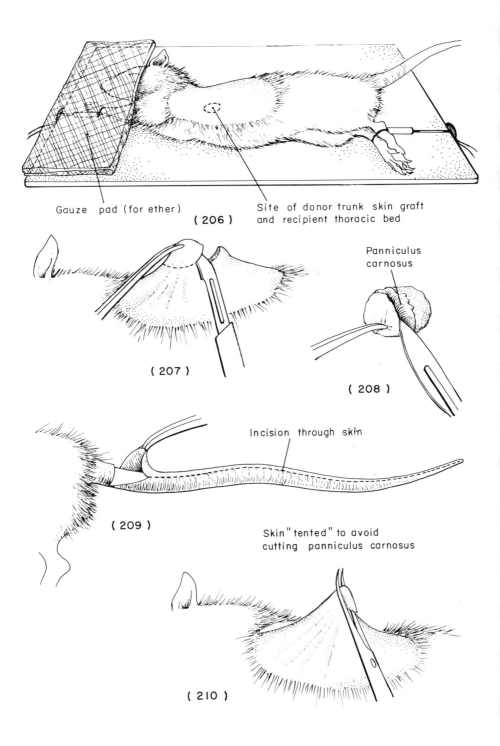

Gauze pad (for ether)

Site of donor trunk skin graft
and recipient thoracic bed

(206)

(207)

Panniculus
carnosus

(208)

Incision through skin

(209)

Skin "tented" to avoid
cutting panniculus carnosus

(210)

dish (on sterile saline-moistened filter paper) hair side down. The panniculus carnosus is then scraped away using the edge of a No. 23 scalpel blade. To do this, the extreme edge of the skin is held down firmly with forceps and the panniculus carnosus at this point is separated from the skin with the scalpel blade. It is then scraped away across the whole piece of skin, which requires a lot of pressure (Fig. 208). The graft is then turned and laid on a clean area of the filter paper hair-side up until required. Skin grafts will remain viable for up to several hours, but they are best used as soon as possible.

35.1.2. Tail Skin Graft

Tail skin provides a complete full thickness graft *per se* as no panniculus carnosus layer is present. This skin can be prepared for bedding into the thoracic area of the recipient or into a bed in the recipient's tail.

35.1.2.1. Tail graft for thoracic skin bed. The tail of a freshly killed rat is thoroughly cleaned with an antiseptic solution and plasticiser is rubbed in and allowed to dry. The skin is then completely cut around the base of the tail and on one side along about three-quarters of the length of the tail with a No. 23 scalpel blade. The last quarter piece of the whole tail is cut off since the skin here is too narrow to be of any use. With toothed forceps the skin is stripped off the tail (Fig. 209) and the under surface is gently scraped with the scalpel blade to remove blood and ligaments. The skin is cut in two and placed on saline-moistened filter paper, hair side up. Rectangular pieces of skin about 15 × 8 mm are used for grafting.

35.1.2.2. Tail graft for tail skin bed. A circular cut is made in the prepared skin of the tail of an anaesthetised or freshly killed rat using a 6 mm diameter sharp sterile stainless steel punch (e.g. a cork borer). (An anaesthetised rat is used only if it is also to be employed subsequently as the recipient of a graft.) The circle of skin so formed is removed from the underlying fascia with forceps and placed on moistened filter

---

Fig. 206. Method of positioning of the rat for skin grafting and showing area for both removal and bedding of the graft. *Fig. 207.* Cutting a circular piece of trunk skin for grafting. *Fig. 208.* Scraping away the panniculus carnosus from the full-thickness graft. *Fig. 209.* Removing the skin from the tail. *Fig. 210.* Preparing a bed in the thoracic skin for the graft, note that the skin is "tented" while making the cut so as to preserve the integrity of the panniculus carnosus and the blood vessels.

paper. It is used after lightly scoring the dermal surface with a scalpel blade, which improves subsequent vascularisation.

35.2. Preparation of a Graft Bed on the Recipient

35.2.1. Thoracic Skin Bed

The anaesthetised rat is laid on its side and its fore and hind legs stretched out to produce slight tension in the thoracic skin. The hair of the lateral thoracic skin is removed and the area is disinfected and plasticised. The skin near the distal part of the thorax is grasped in the blades of a small pair of curved scissors moistened with sterile saline. The skin, which has not been cut at this point, is raised upwards and the scissors then cut off a piece of skin (Fig. 210). Snipping off the skin while it is under vertical tension ensures that only the skin is removed leaving behind the panniculus carnosus and the blood vessels which run through it. The panniculus carnosus and the blood vessels should not be cut, but slight errors of judgement are not detrimental to the outcome of the subsequent grafting. Further cuts are made adjacent to the first, progressing anteriorly, and similar cuts are made above and below these until a roughly circular bed (for thoracic skin grafts) or a rectangular bed (for tail skin grafts) is formed, of an approximate size to accommodate the graft. The graft bed is dried with gauze, ensuring that there is no bleeding in the bed, and the donor skin graft is placed on it. The graft is gently stretched and then trimmed *in situ* with the small curved scissors to fit the bed. A good, but not necessarily an exact fit is required. However, the graft must at no point overlap the graft bed. The graft is covered with a gauze pad and pressed down for a few seconds to ensure that it is bedded in. A square of tulle gras (or Melolin) is placed over the graft and gently pressed against it and the surrounding skin with forceps, care being taken not to dislodge the graft. The animal is then picked up carefully by an assistant while the investigator places several turns of wetted plaster of Paris bandage round the tulle gras or Melolin overlying the graft, and around the thorax. The bandage should fit reasonably tightly round the animal. Again, care must be taken not to dislodge the graft.

If the donor graft is from an animal with the same hair colouring as the recipient, it should be placed in the graft bed so that its hair growth will be in the opposite direction to that of the recipient animal. This is for the purpose of subsequent identification of the graft should it be accepted by the recipient. Animals should be kept singly in cages until the plaster cast is removed on day 9, otherwise they will remove each other's casts.

35.2.2. Tail Skin Bed

The steel punch used to obtain the tail skin grafts (see above) can also be used to produce a bed in the tail skin of the recipient for the donor grafts which will then fit precisely. The grafts are covered with gauze, and are bedded in with finger pressure for several seconds. Because the grafts are small, several grafts of the same or different origin (e.g. autochthonous, syngeneic, allogeneic, xenogeneic) can be placed along the one tail if desired. The precaution mentioned above, concerning similar coloured donor and recipient animals, also applies here.

Protection of the tail graft can take two forms.

(i) The graft is bound firmly with non-allergic adhesive tape and the whole graft site (if more than one graft is involved) is covered with several layers of the same tape.

(ii) A glass centrifuge tube, a little larger than the diameter of the tail and about half the length of the tail, is cut just above the closed end to make an open-ended cylinder. The diameters of both ends are narrowed slightly by placing in a Bunsen flame. This ensures that only the ends of the tube rest on the tail, and not the whole length. The tube is pushed over the tail and over the bedded-in but naked graft, to rest snugly against the buttocks. A Michel skin clip is placed firmly in the tail at the distal end of the glass tube to prevent the tube from slipping off. Alternatively several turns of adhesive tape can be wound round the end of the tube and around the tail. The glass tube prevents the rat from flexing its tail which, if allowed, could dislodge the graft.

A number of interesting variations in technique are available. If the graft cannot be used immediately, temporary storage of the donor skin at 4°C for up to about 3 days before grafting may improve graft survival. Skin grafts can be sprayed with a 2 s burst of aerosol plastic dressing (e.g. Op-Site—see Equipment Index) to "fix" them in place before being covered with plaster of Paris, adhesive tape, or a glass tube in the case of tail grafts. However, there is evidence that some of these procedures can artificially prolong skin graft survival in donor–recipient combinations which show weak genetic histocompatibility differences. Their use, therefore, should be considered carefully and avoided if possible.

As an alternative to tulle gras (or Melolin) and plaster of Paris for thoracic graft protection, a narrow strip of medicated sticky tape dressing (e.g. Elastoplast) can be placed over the graft and the graft protected by several turns of elasticated adhesive bandage (e.g. Flexoplast). Less stress is caused to the animal but the adhesive material is more susceptible to damage by the rat's teeth and may need to be repaired at intervals.

In some instances a percentage of grafts may be found to have slipped from their graft beds. If this is a general problem for the investigator, then it is perfectly feasible to suture the grafts in place initially. A good procedure is to place two sutures in the graft at opposite ends to each other using a small, 10 mm half-circle atraumatic cutting needle bonded to 6/0 suture material (e.g. plain or chromic catgut). Over the graft are now placed, at right angles to each other, two narrow strips of sterile adhesive skin closure tape (e.g. Ethistrip—see Equipment Index). The grafts are then protected with tulle gras (or Melolin) and plaster of Paris or with an adhesive tape bandage, as above.

It is possible to graft areas of skin greater than 2.5 cm in diameter. For example the whole of the back skin of a rat can be removed with scissors as a full thickness graft, and placed on a similarly prepared bed (i.e. no panniculus carnosus present) in the recipient animal. The graft with its panniculus carnosus removed must be completely sutured in place using interrupted stitches or skin clips. For protection of this graft a light dressing can be used.

All skin grafts, irrespective of the genetic make-up of the donor and recipient animals, "heal-in" by about day 6 (N.B. day of grafting = day 1). After this time the fate of the graft is dependent on the extent of the genetic disparity between the two animals. Protective dressings are usually removed on day 9, and removal must be done very carefully so as not to damage or lift the graft which can stick to the dressing. The graft is then observed at regular intervals, daily at first, to see whether the graft has "taken" or is being rejected by the recipient animal. Total rejection is scored when the graft becomes brown, scabbed and often concave (i.e. mummified). However, rejection may take 24 h or more, depending on the treatment the recipient is receiving or on the strength of the histocompatibility difference between donor and recipient. It is often possible to score the rejection process in stages (e.g. 25%, 50% rejected etc.). Rejected grafts are eventually replaced by the recipient's own skin. In grafts that survive for any length of time, the original epidermis sloughs off leaving a new white surface underneath. Healthy grafts will blanch when pressed hard, and recover their colour as blood flows back into the tissue. In general, in donor–recipient combinations which show genetic disparity at the major histocompatibility loci (i.e. RTI), grafts are rejected by day 10 or 11. Where weak histocompatibility is involved only (e.g. male skin on female recipient, or vice versa, in an inbred strain) rejection occurs later, and indeed may not occur for a year or more.

36. SPLENECTOMY

The rat is anaesthetised and laid on its back with its tail towards the investigator. The abdominal cavity is exposed through a midline incision extending posteriorly for about half of the abdomen (Fig. 211).

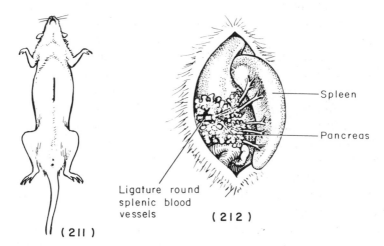

Fig. 211. Site of incision for splenectomy. Fig. 212. Placing of ligature round the splenic blood vessels before removal of the spleen.

Alternatively a dorsoventral incision can be made near to the costal border of the thorax, on the right of the investigator. The prominent splenic vein with an accompanying small artery courses through the pancreatic tissue and splits into several small vessels just before entering the spleen. A ligature or staple is placed around the splenic blood vessels just before they divide; they are severed distal to the ligature and the spleen is removed before closing the incision (Fig. 212).

No special measures are required other than the provision of high standards of postoperative care and rats remain healthy provided they are not infected with *Bartonella muris*, a blood parasite which produces a fatal macrocytic anaemia in splenectomised animals. If necessary, infection can be controlled with an i.m. injection of Neoarsphenamine at a dose of 140 mg/kg body weight.

37. THYMECTOMY

37.1. Adult Rat

The anaesthetised rat is placed on its back with its head towards the investigator. A midline incision is made in the skin from the base of the neck posteriorly over the thorax. The length of the incision should be 2–2.5 cm (see Fig. 213). The connective tissue between the two submandibular salivary glands, which are exposed by the incision, is grasped with two pairs of forceps which are then pulled sideways, away from each other. This separates the two glands. In older rats the connective tissue may have to be cut as it is rather tough. This procedure exposes the sternohyoid muscle which overlies the trachea. The muscle is in two halves kept together by connective tissue, and the two halves are separated by running the points of a pair of forceps up and down the centre of the muscle while pressing down quite hard. This causes the connective tissue to tear and the medial plane between the two halves to be revealed. These can now be retracted to either side. The thorax is opened by cutting through the sternal manubrium and pectoral muscles in the midline with blunt-ended scissors (Figs 214 and 215). It is important that the cut is made long enough to give good retraction of both halves of the manubrium, and should be 1–1.5 cm in length so that the manubrium is completely transected down to the first sternebra

(Fig. 216). The investigator should find the two halves of the manubrium quite pliable if the transection has been made correctly. However, too long a cut will produce a pneumothorax, and care must therefore be taken. The two halves of the sternohyoid muscle are further separated down to the cut end of the manubrium, and the top of the thymus gland should now be clearly visible (Fig. 217). Connective tissue between the top and sides of the anterior end of the gland is cleared with forceps and the two halves of the sternohyoid muscle and of the manubrium are retracted as widely as possible on either side.

The thymus gland is a bilobed, roughly heart-shaped structure with its broad end towards the investigator. Each lobe is more or less separate, but held closely together by a connective tissue covering. For its removal, the gland is picked up with forceps and a second pair of forceps is pushed under it and opened to tear the connective tissue attachments. The connective tissue at the sides of the gland is torn with forceps or snipped with scissors (Fig. 218). This is an important step which frees the anterior part of the gland and must be done with care because of the proximity of major blood vessels. The tip of the gland is now held in the middle by a pair of curved medium broad artery or Spencer–Wells forceps. These are self-holding forceps and should be broad enough to hold a piece of each lobe if positioned correctly. The gland is pulled outwards and upwards gently and the connective tissue covering the gland is pushed back towards the heart (Fig. 219). This is done with a thin wooden or plastic rod, around one end of which is wound a small piece of cotton wool. This bulb of cotton wool must be quite slim so that the applicator can be pushed along the sides and the under surface of the gland. The cotton-tipped applicator is then pushed into the thoracic cavity on top of the posterior end of the gland, and by a slight scooping and pressing motion of the applicator towards the investigator, together with a gentle pull on the gland by means of the artery forceps, the thymus gland is removed (Fig. 220). As the gland tears away from the rest of the connective tissue there may be a tendency for the gland to break up. To prevent this the cotton-tipped applicator can be turned on its side and pressed down hard across the posterior part of the gland and rolled towards the investigator. This "gathers up" the gland and reduces the amount of pull required from the artery forceps (Fig. 221).

Occasionally the gland may separate into its two lobes which must therefore be removed separately. Any bleeding that occurs during or after removal of the gland, even rather copious bleeding which sometimes

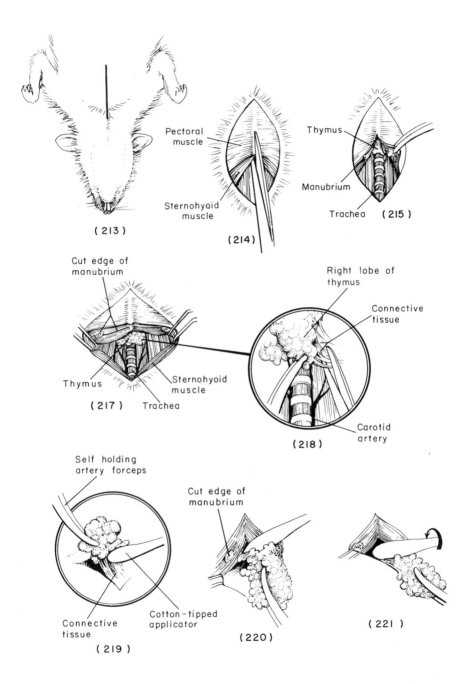

Pectoral muscle

Sternohyoid muscle

(214)

Thymus

Manubrium

Trachea   (215)

(213)

Cut edge of manubrium

Right lobe of thymus

Connective tissue

Thymus

Sternohyoid muscle

(217)   Trachea

Carotid artery

(218)

Self holding artery forceps

Cut edge of manubrium

Connective tissue

Cotton-tipped applicator

(220)

(221)

(219)

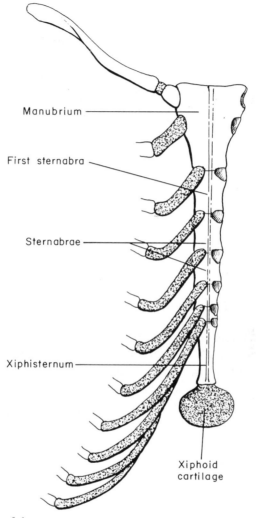

Fig. 216. Anatomy of the sternum.

Fig. 213. Site of incision for thymectomy. *Figs 214, 215*. Cutting through the sternal manubrium and pectoral muscles. *Fig. 217*. Exposure of the top of the thymus gland. The two halves of the sternohyoid muscle and the cut manubrium are widely retracted. *Fig. 218*. Cutting (or tearing) the connective tissue at the sides of the thymus gland. *Fig. 219*. Clearing the connective tissue under and around the thymus gland with a cotton-tipped applicator. *Figs 220, 221*. Pulling out the thymus gland with the aid of a slim cotton-tipped applicator and artery forceps.

occurs in older rats, can be stopped with pressure from cotton-tipped applicators, providing the major blood vessels are not involved. Immediately after removal of the gland the animal frequently shows signs of respiratory distress, sometimes due to the inadvertent production of a pneumothorax. To prevent or stop this it is important to bring the cut ends of the sternum together very quickly and place a short continuous suture through the overlying pectoral muscles to close the thorax. Rats recover quickly from the operation and show excellent survival. No special treatment is required other than a high standard of postoperative care.

37.2. Neonatal Rat (adapted from Hard, 1975)

The very small size of the animal necessitates the use of small instruments, e.g. iris scissors and iris forceps. Sterile instruments are placed on sterile drapes and sterile gloves and a face mask should be worn. The operating board is swabbed with antiseptic before the operation.

Neonatal rats are anaesthetised using methoxyflurane, or by hypothermia, produced by placing the animal in a refrigerator cooled to 5°C for about 25 min until all movement ceases. The anaesthetised rat is fixed to the operating board by adhesive tape placed across the chin and the hind and fore limbs, and its thorax is arched by placing a small sterile roll of cotton gauze or foam rubber beneath its neck (Fig. 222). The basic approach to exposure of the thymus gland is identical to that in the adult rat, bearing in mind the reduced size and delicate nature of the tissues of the neonate. Cutting through the sternum should proceed through to the end of the first sternebra, a distance of about 5 mm. However, too long a cut into the sternum will lead to excessive and possibly fatal haemorrhage. The cut ends of the sternum should be kept apart by an opened pair of iris forceps acting as a retractor, so that after separation of the two halves of the underlying sternohyoid muscle the thymus gland is wholly exposed, but only just so (Fig. 223). This final step in the exposure of the gland and all subsequent procedures are carried out under a binocular microscope.

To remove the gland, which is multilobed, each lobe is picked up separately with a short piece of sterile polythene tubing attached to a water suction pump. The tubing has an i.d. of 1 mm and is attached

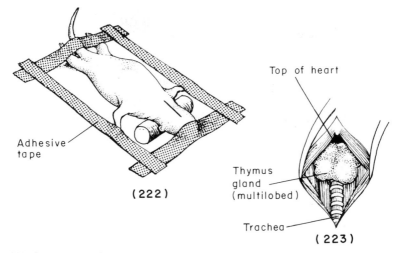

Fig. 222. Positioning of a neonatal rat with adhesive tape; the site of the incision for thymectomy is shown (adapted from Hard, 1975). Fig. 223. The multilobed neonatal thymus gland, note that the top of the heart should be only just visible if the length of the incision has been made correctly.

to a piece of glass tubing containing an idle air hole by which the force of the vacuum can be controlled depending on how much of the hole is blocked off by a covering finger. Since the gland is fairly well attached by connective tissue in the region of the heart, some blunt dissection will have to be carried out to free each lobe during the suction process.

Before closing the skin incision (there is no need to suture together the cut ends of the sternum), a little benzylpenicillin solution (see 58) should be dropped into the wound as a prophylactic measure. The skin is closed with interrupted sutures of 5/0 suture material using a small atraumatic 3/8 circle needle. The animal is placed in a hot box or incubator at 37°C for 2 hours before being returned to its mother.

To ensure a high survival rate and a successful operation a number of conditions have to be observed.

(i) Fatal postoperative infection and the "runting syndrome" are reduced to a very small number of animals if a strictly sterile technique is used throughout.

(ii) A considerably reduced incidence of cannibalism of the neonate by the mother postoperatively can be achieved by tranquillising the mother (e.g. with diazepam, 1 mg/kg i.p.) before removal of the neonate from the cage. It is also helpful to smear urine or faeces from the mother, collected at the time of the tranquillising injection, over the neonate before returning it to its still tranquillised parent.

(iii) The use of a binocular microscope is essential to check for the removal of residual pieces of thymic tissue which sometimes remain attached to the overlying fascia after removal of the major part of the gland.

38. THYROIDECTOMY (THYRO-PARATHYROIDECTOMY)

The paired thyroid glands are small pink organs, one on either side of the trachea just below the larynx. They are connected across the ventral aspect of the trachea by a thin band of tissue, the isthmus. The minute parathyroid glands are embedded in the anterior part of each thyroid gland and cannot be separated satisfactorily from them. The operation is technically, therefore, a thyro-parathyroidectomy. Removal of the thyroid is most satisfactorily carried out with the use of a single-edged iridectomy scalpel (e.g. Graefes cataract knife), or if unobtainable, a dissecting needle, in addition to the basic instruments.

The rat is laid on its back with its head or tail towards the investigator. A midline skin incision is made along the length of the neck from its base to just below the point of the lower jaw (Fig. 224). The subcutaneous fat and connective tissue are cleared and the salivary glands are separated and retracted laterally by grasping the connective tissue between them with two pairs of forceps and then pulling them sideways from each other. The two halves of the sternohyoid muscle which are now exposed are separated and retracted laterally (see 37). Along the dorsolateral aspect of each lobe of the thyroid runs the thyrohyoid muscle (Fig. 225). This is separated from the gland by blunt dissection and retracted along with the sternohyoid muscle.

The isthmus is grasped with forceps (small-toothed iris forceps are useful) and cut in the midline with the iridectomy scalpel. The piece of isthmus tissue connected to the right gland (left of the investigator) is grasped and gently pulled sideways to place the gland under lateral

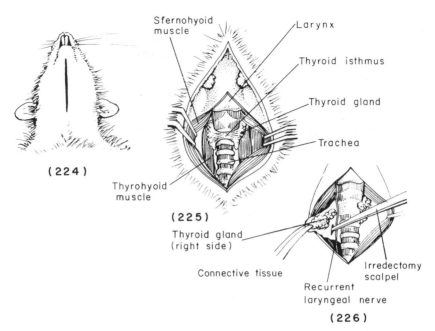

Fig. 224. Site of incision for thyro-parathyroidectomy. Fig. 225. The thyroid gland and its anatomical relationships. Fig. 226. Removal of the thyroid gland (right side), note the position of the recurrent laryngeal nerve, which must not be cut.

tension. The connective tissue between the gland and the trachea is carefully cleared by repeatedly pushing the gland away from the trachea with the flat side of the scalpel, working progressively down the gland. As the dorsal edge of the gland is approached, the investigator must proceed with extreme caution since along it runs the recurrent laryngeal nerve (Fig. 226). Severing the nerve can cause paralysis of the larynx, and the respiratory impairment can be fatal. The nerve will be seen as a fine white strand loosely attached to the edge of the thyroid by connective tissue. When found, the nerve should be pushed away from the gland towards the trachea, again with the flat side of the scalpel. The gland is finally severed from its vascular connections anteriorly and posteriorly, and removed. Any bleeding that occurs during and after removal of the gland is controlled with small cotton wool pledgets. The procedure for removal of the left gland is identical, but it is easier to work on this gland if the rat is repositioned left (head) and right of the investigator. Because of the lateral tension continuously placed on the gland, the piece of isthmus tissue held may break away. In this case the gland itself will have to be held, which will not matter greatly as the thyroid is enclosed in a tough connective tissue capsule. Before closing the skin incision, retractors and cotton wool pledgets are removed and the muscles allowed to return to their normal positions. No suturing of the sternohyoid muscle is required.

Since the parathyroid glands are removed along with the thyroid, young rats may experience tetanic convulsions because of the loss of control over calcium homeostasis. To avoid this, young rats should be given 2% calcium lactate to drink *ad libitum* until calcium homeostasis is restored by other factors taking over this process in the body. This state is reached postoperatively after about 10 days. The rat also has adventitious parathyroids which can usually cope in the older animal.

Chapter Six

MISCELLANEOUS TECHNIQUES

39. RECORDING BODY TEMPERATURE

Body temperature can be recorded simply and easily using an ordinary mercury thermometer but it is much more convenient to use an electronic thermometer with its temperature probe placed in the rectum (Fig. 69). A number of different designs are available, a probe size of 2–4 mm is suitable for adult (200 g) rats, 1–2 mm probes are suitable for use in younger animals. For continuous recording of temperature the electronic thermometer can be interfaced to a chart recorder or to a microcomputer. Thermistor temperature probes can also be implanted subcutaneously and connected to a recording device by means of a harness and tether system, as described for vascular cannulae (Section 17.3). Recording of temperature using an implanted sensor and a radiotelemetry device is also possible (Meindl et al., 1986).

40. RECORDING BLOOD PRESSURE

Arterial blood pressure can be recorded either directly, by cannulation of a convenient artery, or indirectly using the tail cuff technique.

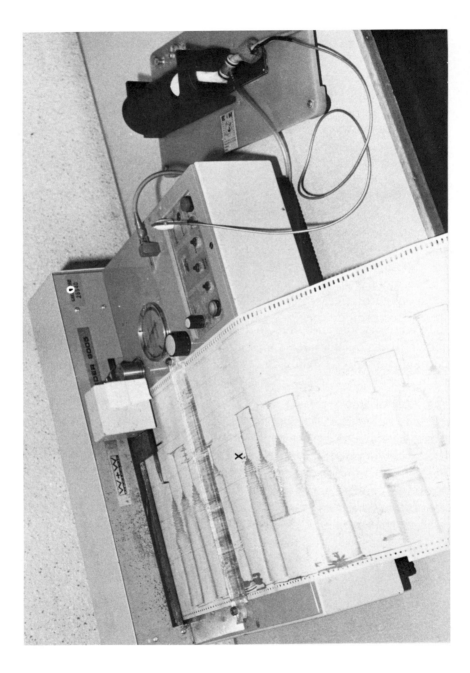

40.1. Indirect Method

This technique makes use of a special inflatable cuff and a pulse detector, both of which fit over the tail and are connected to a blood pressure recorder. These are all available commercially (see Equipment Index). The apparatus is shown in use in Fig. 227.

Initially the rat is warmed in a 37°C hot box (see 2.15.1.1) for about 15 min (without warming the pulse may not be detected by the recorder). The rat is placed in a restraining apparatus which can be individually heated if required. The inflatable cuff containing a disposable rubber lining is pushed onto the tail first, followed by the pulse detector cuff which should fit snugly. The pulse is first tested and, if good, the recorder is started and the pulse rate recorded. The cuff is then inflated automatically until its pressure against the tail restricts the blood flow to such an extent that a pulse can just no longer be detected. This pressure is equal to the systolic blood pressure. Usually the mean of four or more readings is obtained for each animal. Heart rate can also be measured by the recorder and is usually read at the end of the blood pressure recordings when the rat has become more settled.

40.2. Direct Method

The basis of this approach is as follows (see also 22). The left carotid artery is catheterised such that the tip of the plastic catheter lies just inside the aortic arch. The catheter must be passed subcutaneously and exteriorised dorsally at the neck with about 2 cm showing. The catheter is filled prior to insertion with a dilute heparin–saline solution (20 units heparin/ml) and is stoppered with a pin. The rat is allowed to recover from the catheterisation process and is used after 3 or more days. (The carotid catheter should be flushed through daily to maintain patency and also just prior to blood pressure measurement.) At this time a second heparin–saline-filled catheter is connected to the first with continuity of the heparin–saline solutions, and the other end of the

Fig. 227. Commercially available apparatus for recording blood pressure by the indirect method; the rat restrainer can be heated electrically if required. Note the decrease in pulse beat on the recording paper as the tail cuff is inflated. At point X no pulse is obtained and the reading here equals the systolic blood pressure.

second catheter is connected to a transducer. Alternatively, the carotid catheter can be connected to the transducer and blood pressure (BP) measurement system immediately after surgery using the harness and tether restraint system (see 17.3). The pulsating movement of blood in the carotid artery causes pressure changes in the heparin–saline solution which are transmitted to the transducer. The transducer can be connected either to a chart recorder or to an oscilloscope-type display, or interfaced to a microcomputer via a suitable AD convertor (see Equipment Index). Relatively inexpensive transducers, designed for single use in human patients, can be used successfully to monitor blood pressure in rats. These disposable transducers can be reused for a considerable length of time for non-recovery experiments. They can also be resterilised using ethylene oxide, but must be flushed thoroughly with sterile saline before reuse to ensure that no gas remains in the transducer.

41. *IN VIVO* PERFUSION TECHNIQUES

Only a brief outline of this subject will be given here since in recent years this has become a broad and specialised field (see Ross, 1972). Perfusion can be carried out with the organ remaining either *in situ* throughout the perfusion or the organ can be removed and placed into a special *in vitro* environment during which perfusion continues. In the first case, perfusion is often of a relatively simple nature and the method is often used in some areas of biochemistry and for histological fixation. When it is necessary to preserve organs for more complex procedures, e.g. transplantation, the tissues are usually kept isolated during the major part of the perfusion process. In these circumstances it is necessary to choose the best type of perfusion medium for a particular organ, to control the gas content and temperature of the perfusion medium, to choose the best apparatus to deliver the medium to the organ concerned, and to maintain adequate circulation. The following brief considerations apply to perfusion of organs.

(i) Perfusion medium. This will depend on the study to be made. In the simplest case physiological saline may suffice, but it is usually expedient to use buffered solutions of a more complex composition. Of these, the commonest perfusion media are probably Ringer solutions and Krebs–Henseleit buffer.

(ii) pH. This may also vary with the requirements of the study. In most cases a pH of about 7.2 is used.

(iii) Gas content and temperature control. In the simplest case no provision is made to control the oxygen or carbon dioxide content of the perfusion medium, and the temperature is usually that of room temperature. In more specialised studies the medium is gassed using oxygenators of which there are types of various complexity. The temperature of the medium can be controlled using a water jacketed apparatus or a perfusion cabinet maintained at the required temperature.

(iv) Perfusion apparatus. In the simplest case perfusion can be carried out using a syringe and needle. However in many cases it is preferable to catheterise the appropriate blood vessels and to connect the catheter to a reservoir connected to a pump capable of producing either a pulsatile or non-pulsatile flow, depending on requirements. This ensures a continuous and steady flow of medium, and by incorporating a device to measure pressure, the medium can be delivered at a pressure associated with the normal pressure of blood flowing through a particular organ in the intact animal. The perfusion itself can be a "once-through" operation or can be arranged so that a preselected amount of perfusion medium is recirculated through the organ. A simple home-made apparatus using a blow-ball to supply the propulsive force (or a sphygmomanometer can be used as an alternative) is shown in Fig. 228. A simple perfusion procedure to illustrate methodology is described below.

41.1. Liver Perfusion *in vivo*

41.1.1. *"Once-through" Procedure*

The anaesthetised rat is placed on its back with its tail towards the investigator. The abdominal cavity is opened with a V-cut (see 3.3.3) and the abdominal flap reflected onto the chest wall. The gut is pulled over to the investigator's right side and the large hepatic portal vein is located. Any small venous off-shoots from the portal vein are ligated if desired. The vein is brought under tension by straddling it with the index and middle fingers (see 2.15). It is then entered using a 21G

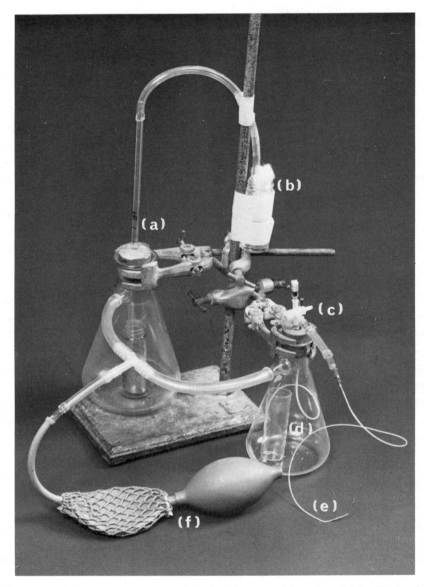

Fig. 228. Perfusion apparatus prepared in the laboratory: (a) graduated tube dipping into mercury in universal container (i.e. a manometer); (b) container to collect any overflow of mercury, (c) three-way stopcock, (d) container for infusion medium, (e) polythene tubing for catheterisation, (f) blowball with air reservoir.

needle connected to a 50 ml syringe filled with perfusion medium (Fig. 229). Immediately on entering the vein, the perfusion is started slowly and the posterior vena cava is cut with scissors between the liver and the diaphragm to allow the perfusate to escape after it has passed through the liver. As the blood is displaced from the liver, the latter will become progressively paler in colour (providing the perfusion medium is not itself blood). To facilitate removal of blood from the peripheral parts of the liver lobes, it may, in some cases, be necessary to massage them gently between the fingers. Subsequent to the perfusion of about 50 ml of medium, the liver is removed by gripping the diaphragm with forceps, and cutting round it to free the liver and then cutting away the remnants of the diaphragm from the liver. Liver tissue is thus obtained which has not been damaged by direct handling with forceps.

41.1.2. Recirculation Technique

The hepatic portal vein is exposed as explained above. Two ligatures are placed loosely (not tied as yet) around the vein with 3–4 mm between the ligatures. A catheter (i.d. 1.14 mm, o.d. 1.57 mm) is prepared and connected to a recirculating pump and a reservoir. An arrangement incorporating temperature control and gassing of the perfusion medium with 95% air–5% carbon dioxide from a gas cylinder is shown in Fig. 230. The posterior ligature is now tied tightly. The free ends of the anterior ligature are pulled upwards by an assistant to constrict the vein momentarily and prevent backflow of blood. The vein is semitransected between the two ligatures, and the catheter passed into the vein and a little beyond the anterior ligature. This is now tied tightly around the vein and the catheter to keep it in place. The perfusion should be started slowly. The thoracic cavity is now opened quickly and the anterior vena cava catheterised and ligated. (Alternatively it may be simpler to catheterise the posterior vena cava.) This catheter is connected to the reservoir to return the perfusion medium after it has passed through the liver. Perfusion is then carried out with the animal lying below the reservoir so that 8–20 cm of water pressure is achieved for the medium. Ideally the perfusion flow rate should be 10–20 ml/min with the medium maintained at 37°C.

A number of points should be noted concerned with liver perfusion and which, with suitable modification, are applicable to organ perfusion

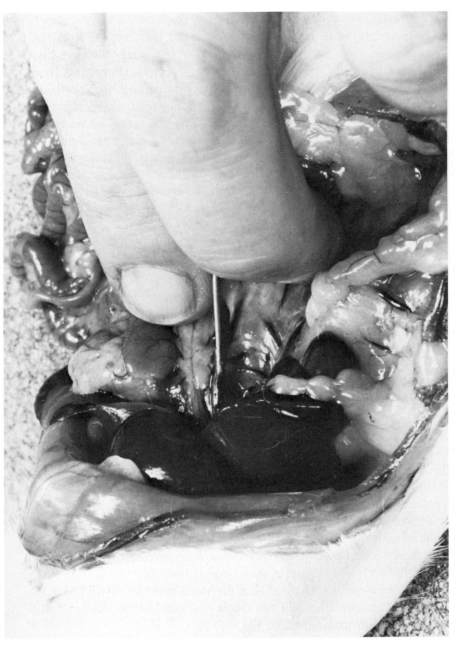

Fig. 229. Liver perfusion via the hepatic portal vein using a syringe and needle.

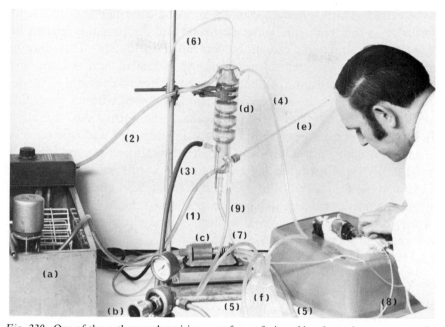

Fig. 230. One of the authors catheterising a rat for perfusion. Also shown is an apparatus for continuously circulating the perfusion medium and controlling its gas content and temperature: (a) heated water bath and pump, (b) 95% air: 5% carbon dioxide gas cylinder, (c) circulating pump, (d) water jacket, (e) thermometer, (f) reservoir with perfusion medium. Tubing and directions of flow: (1) water flow to jacket, (2) water flow from jacket, (3) gas input to column, (4) gas overflow to reservoir, (5) reservoir to pump, (6) pump to column, (7) column to rat, (8) flow (by gravity) from rat to reservoir, (9) overflow of medium to reservoir to maintain level in column at the point of overflow. The rat should be kept at 8–20 cm below this point.

in general. First, the principal blood vessel supplying an organ should be catheterised preferentially. Where this vessel is too small, then catheterisation of some other connected vessel, with appropriate placing of ligatures so that the perfusion medium is directed only to the organ concerned, will suffice. For example, to perfuse the kidney whose renal artery is extremely small (≤ 1 mm), the dorsal aorta is first catheterised above, and then ligated below the origin of the renal artery. Other arteries arising from the aorta between the ligature and the site of entry of the catheter are also ligated. Perfusion will now ensure that all the perfusion medium is channelled into the kidney. Second, if it is desired to recirculate the medium or to collect the medium after it has passed through the organ, then the appropriate vein from that organ must be catheterised. Third, one disadvantage of using a syringe and needle for perfusion is that both the rate of perfusion and the pressure applied for propelling the medium cannot be controlled accurately. This may be critical, at least for some histological and ultrastructural studies. For example, only perfusion of the kidney at a steady pressure of about 180 mmHg, the maximum pressure that can be used without exceeding the recorded peak systolic blood pressure of the normal rat, will allow the lumen of the kidney tubules to remain open and undistorted. For a "once-through" procedure, the simple apparatus shown in Fig. 228 could easily be substituted for a syringe and needle to achieve a more controlled flow. Lastly, if it is desired to perfuse the organ in the isolated state, catheterisation can be performed more conveniently *in vivo* and the organ then severed from its *in situ* connections and placed into the appropriate isolation environment.

42. TUMOUR TRANSPLANTATION

Tumours vary in consistency from solid forms such as fibrosarcomas, to fluid tumours consisting of a cell suspension, exemplified by an ascites tumour. In addition, by suitable treatment (see 44.7) solid tumours can be dissociated and single-cell suspensions prepared. Simple injection with a sterile syringe and large-bore needle will suffice to transplant tumour cell suspensions or a fine mash of tumour.

The standard procedure for transplanting solid tumours, which requires the use of sterile technique throughout (see 14), involves the use of a trocar. This is a needle of very large bore (1–3 mm i.d.) containing a manually operated stylus (Fig. 231—see Equipment Index). The procedure is as follows. The tumour is dissected from the rat and all extraneous tissue is removed. It is then cut in half in a Petri dish and rinsed with saline and transferred to a second dish. In many cases the tumour will be found to consist of a central, discoloured, necrotic area surrounded by a white area of viable tumour tissue. Only the latter must be used for transplantation. The white tissue is cut out and transferred to a third Petri dish. Small rectangular pieces of tissue, about 1–3 mm in size, are cut to fit into the small size trocar. Larger sizes can be cut to fit larger size trocar needles. Three or four pieces

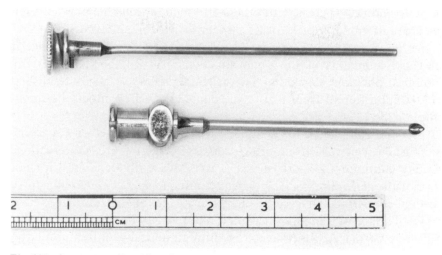

Fig. 231. A trocar needle with stylus.

are forced gently into the bevelled end of the trocar and the stylus inserted part way into the other end, ensuring that the tissue is not forced out.

For the transplantation, the hair of the animal is removed over a small area which is then swabbed with antiseptic. The loaded trocar is then inserted fully subcutaneously into either the conscious or anaesthetised animal, and the pieces of tumour deposited by pushing in the stylus completely. For ease of establishment and optimum growth of tumour, the most favourable sites for tumour deposition are the axillary and inguinal regions because of the abundant blood supply to these regions.

Although a trocar facilitates the transplantation procedure, it can be carried out quite adequately under anaesthesia by making a small incision in the skin. A subcutaneous pocket is prepared by opening a pair of blunt-ended scissors under the skin, and then tumour tissue is deposited in the pocket with a pair of straight forceps. The tissue should be deposited as far from the incision as possible to prevent loss of the tumour pieces, and the incision closed with one or two skin clips or sutures.

The following considerations are relevant to tumour transplantation.

(i) The number of cells in the transplant may be critical for a "take" to be achieved. With some tumours a single cell will give rise to a tumour nodule, whereas with others 10^6 or more cells may be required (as a rough guide, a piece of tissue $1mm^3$ will contain 10^6 cells).

(ii) The most successful establishment of a transplant takes place if the host animal is of the same strain and sex as the donor. Many tumours, however, depending on their antigenicity, can be established in a strain different from that of the donor rat. This can only be found empirically.

(iii) The time taken for a tumour transplant to "take", i.e. become established, can vary from a few days to a year or more. Observations should continue therefore for at least 1 year. In general, the first transplant generation of the primary tumour takes much longer to become established than subsequent transplants of that tumour.

(iv) The site of tumour transplant can be diverse. For example, tumour cells can be placed subcutaneously, intraperitoneally, intramuscularly, intravenously, into the eye, under organ capsules, into organs and in practically any other site. The establishment of a tumour, however, may depend on the site that is used.

43. ANTIBODY PRODUCTION

The general techniques for producing antibody to a particular antigen in the rat are the same as those used for all other species. The most widely used method is based on the use of complete or incomplete Freund's adjuvant. However, because Freund's complete adjuvant invariably causes severe inflammation manifesting as lumps, or abscesses which in some cases may rupture, recommendations have appeared in the literature for the use of other, less debilitating or distressing adjuvant mixtures. Examples of these (see Equipment Index) are the RIBI adjuvant (a mixture of trehalose dimycolate, monophosphoryl lipid A and cell wall skeleton in a squalene-in water emulsion), montanide ISA50 and 70 emulsions, simple formulations of mannide oleate in mineral oil and Titermax™ (consisting mainly of squalene plus a block co-polymer adjuvant). These alternative adjuvants have given good results when used with some antigens but none has equalled Freund's complete adjuvant in the universality of its application (see Johnston et al., 1991). Nevertheless because of their potential in causing less suffering and pain to the animal, they should be used where there is reason to suspect that they will produce an adequate antibody response to a particular antigen. However, the following description will be for the use of Freund's adjuvant. Complete Freund's adjuvant is a mixture of mineral oils and between 1 and 10 mg/ml of heat-killed *Mycobacterium tuberculosis*. Incomplete Freund's adjuvant does not contain the *Mycobacterium*. Freund's adjuvant with and without *M. tuberculosis* is cheap and readily available commercially. Its principal function is twofold. First, it induces a local inflammatory response (very vigorous if the complete form is used) which attracts an increased number of white blood cells and macrophages into the area. This ensures that maximum contact is made between the antigen carried in the adjuvant and the body's immunological cells which produce antibodies or display other appropriate immunological responses. Second, since the antigen solution is prepared as a water-in-oil emulsion (see below), the antigen "locked up" within oil droplets is released very slowly, thereby producing a prolonged and continuous stimulation of the host's immunological system. The bacterial component of complete Freund's adjuvant may in some cases induce abscess formation, which is to be avoided if possible, since this may cause unnecessary distress to the rat. Furthermore, if the abscess bursts, the adjuvant and antigen will be discharged and no longer be effective in stimulating an immune response. Incomplete Freund's adjuvant produces a less marked inflammatory response, and abscessation is less common. It is claimed, however, that the immune response is also reduced. In virtually all instances, a single challenge with antigen and complete Freund's adjuvant, followed by subsequent injections with incomplete Freund's adjuvant, will produce satisfactory results. A number of synthetic adjuvants are currently undergoing evaluation, and it is hoped that these may provide enhanced antigenic stimulation without adverse effects.

To prepare the adjuvant–antigen mixture for injection, equal volumes of adjuvant and aqueous antigen solution are taken up separately in two Luer–Lok syringes. Any air in the syringes is expelled and the two syringes are interconnected via a three-way plastic or metal stopcock (Fig. 232). The antigen solution is expelled forcefully into the adjuvant-containing syringe and this mixture is forced back into the other syringe. This movement of the mixture from one syringe to the other is continued for 100 or so times until the correct consistency of emulsion is produced. To test this, a small drop of the emulsion (use a 25G needle to produce

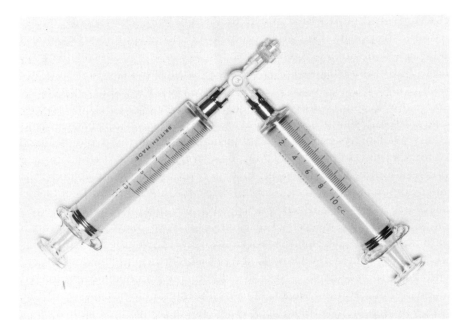

Fig. 232. Two Luer–Lok syringes interconnected via a three-way plastic stopcock.

the drop) is allowed to fall into a beaker of cold water. This is followed by a second, and if necessary a third drop. The first drop usually spreads over the surface of the water but the subsequent drops should stay as discrete white drops showing that the aqueous antigen phase is enclosed entirely within the oil. If no discrete drops are formed, the syringe mixing must be continued until the correct integrity of the emulsion is achieved. If this is not done immunisation may fail.

If a stopcock is not available, emulsion formation can be carried out by mixing the antigen and adjuvant in a tube and then taking this up into a Luer–Lok syringe via a 19G or 21G needle. The mixture is now expelled through a 25G needle. It is taken up once more through the larger needle and expelled through the smaller needle. The operator should carry out this operation in a fume cupboard, while wearing a face mask to avoid inhaling the aerosol which is produced. This recycling operation is repeated until the water test outlined above is successful. A third method of preparing an emulsion is to mix the adjuvant and antigen solution using a vortex mixer (e.g. Whirlimixer) for at least 10 min. Success by this rather simple method, however, is very variable.

The administration of the emulsion is made in small volumes (up to 0.5 ml). There are numerous descriptions of techniques which are claimed to result in superior immune responses, but in virtually all instances no controlled trials have been undertaken. The use of multiple sites of injection, particularly into the muscles and the footpads, is rarely justified. A marked inflammatory response invariably follows administration of adjuvant, and this can cause the rat considerable pain. If a footpad is used as the site of administration, the rat may become lame. Use of more than one footpad will force the animal to use one of the inflamed and painful pads. Similar problems can arise when the intramuscular route is used.

Before attempting to raise antibodies to a specific antigen in the rat, try to establish in a pilot study whether it is necessary to use an adjuvant at all. Some material, particularly particulate antigens such as cells or bacteria, are sufficiently antigenic to produce an effective stimulus without the use of any adjuvant. If an adjuvant is required, use incomplete Freund's, administered at a single subcutaneous site, with a small volume (<0.2 ml) of material. If this regimen does not produce a sufficient response, then use of complete Freund's adjuvant and of additional sites of injection can be contemplated. Since there appears to be no evidence that the use of the footpad produces an enhanced immune response, this site should not be used.

The antibody response to an injection of antigen in complete Freund's adjuvant usually reaches a peak between 4 and 8 weeks after the immunisation, and thereafter remains at a reduced plateau level for several months. A booster dose of antigen, if given after 3–4 weeks, usually increases the antibody response. Ideally a number of booster doses should be given until no further increase in antibody response is obtained, or the antibody titre is adequate.

The serum antibody response should be tested 4–8 weeks after the initial injection, and 7–10 days after each booster injection. Providing only small quantities of blood are taken on each occasion (less than 10% of circulating volume), the rat can be boosted intermittently over a number of months to produce further quantities of serum with maximum antibody activity.

An important question concerning the production of antiserum is how much antigenic material to use. Since this depends on the antigenicity of the material, which for an untested antigen will not be known, the amount has to be empirical. As little as 1 μg of some antigens can elicit a good response. Probably a quantity of 1 mg should be tested initially if nothing is known about the antigen.

44. THE PREPARATION OF SINGLE CELL SUSPENSIONS

Cell suspensions are usually prepared by mechanical or enzymatic means, or by a process of gravity sedimentation. The preparation of various types of cells in suspension is described below.

44.1. Red Blood Cells

Centrifugation of blood (400 times gravity for 10 min at 4°C) collected with the addition of an anticoagulant such as heparin (100 units/ml) or in Alsevers or acid citrate dextrose solutions which prevent the blood clotting, will produce a lower tube component of rbc and a serum supernatant which should be discarded. A fine white layer of white blood cells (wbc) (the buffy coat) which sits on top of the rbc can also be removed. The rbc should be washed three times with a physiological salt solution before use.

44.2. White Blood Cells and Blood Lymphocytes

Two methods are in common use, both dependent on sedimentation.

44.2.1. Method 1

(i) Prepare a 2% solution of dextran (molecular weight 250 000) in phosphate buffered saline (PBS) (this can be sterilised if required by autoclaving at 15 lb/in^2 (3 kg/cm^2), 120°C, for 15 min). Add an equal amount of blood containing an anticoagulant (equals 1% dextran in the final concentration). (N.B. 4% polyvinylpyrrolidone (PVP) can be used instead of 2% dextran, 1 ml of 4% PVP should be added to 4 ml blood containing anticoagulant.)

(ii) Mix gently by inversion, and place at 37°C for 40 min (or leave at room temperature for 1 hour).

(iii) Remove supernatant which contains the wbc.

(iv) Centrifuge supernatant and wash cells three times with PBS. The wbc produced are significantly contaminated with rbc. To remove these, 10 volumes of Tris–NH$_4$Cl, pH 7.2, should be added to a pellet of cells and mixed and left at room temperature for 10 min. The cells should be centrifuged at 150 times gravity for 10 min and washed three times with PBS. Tris–NH$_4$Cl produces gentle lysis of the rbc without affecting the wbc. The wbc obtained contain predominantly lymphocytes and polymorphonuclear cells.

44.2.2. Method 2

Ficoll-Paque (Pharmacia) or Lymphoprep (Nyegaard) can be obtained commercially (see Equipment Index, N.B. similar preparations are marketed by other companies). Both separation fluids are a mixture of the high molecular weight polymer Ficoll 400 and the high density compounds diatrizoate sodium (Ficoll-Paque) or sodium metrizoate (Lymphoprep). The density of both solutions is 1.007 g/cm^3 and they are sterile.

(i) Place 3 ml of separation fluid into a centrifuge tube and carefully layer on top with a Pasteur pipette 4 ml of diluted blood containing anticoagulant and diluted 1:1 with PBS or balanced salt solution (BSS). Mixing must be avoided. Alternatively, if the blood is placed in a syringe connected to a 15.25 cm long 14G needle cannula, it can be expelled slowly at the bottom of the tube, below the separation fluid and this results in an extremely sharp interface and avoids inadvertent mixing.

(ii) Centrifuge the tube for 30 min at 18–20°C at exactly 400 times gravity, measured at the interface between the blood and the separating fluid.

(iii) Remove the white band of cells at the interface with a Pasteur pipette taking care to aspirate as little as possible of the fluids above (containing the platelets) and below the band of cells.

(iv) Wash cells in BSS three times, centrifuging at 150 times gravity for 10 min at 4°C. Resuspend in BSS or PBS.

The cells obtained are virtually pure lymphocytes, however only about 50% of the lymphocytes in the original blood are recovered by this procedure. If required, the wbc fraction from Method 1 (in 5 ml BSS) can be layered onto separation fluid (3 ml), and the lymphocytes isolated as in (ii).

44.3. Lymph Node Cells

Mesenteric, axillary, brachial and cervical lymph nodes are among the largest, and will give a good supply of lymphocytes. Single cell suspensions are obtained by mechanical means with a viability of 80–95% using the following procedure:

(i)   Place the nodes in a little BSS or PBS.
(ii) Tease apart using two pairs of fine forceps, or press between a pair of smooth broad-tipped forceps. This releases many of the cells into the medium.
(iii) Place the residual tissue onto stainless steel wire gauze or into a wire gauze sieve with mesh No. 100 and having apertures of 150 μm (see Equipment Index). Using the flat rubber end of the plunger from a 2 or 5 ml disposable syringe, or a rubber bung or the flattened end of a glass rod, press the tissue gently against the wire mesh. Wash the cells through with BSS or PBS.
(iv) Centrifuge the cells at 150 times gravity for 10 min at 4°C and wash the cells three times with cold BSS or PBS.
(v)   Resuspend the cells.

During the washing procedure a certain amount of clumping of dead cells may occur. The clumps can be removed either by passing through wire mesh or through nylon bolting cloth with mesh No. 100 (see Equipment Index). From the mesenteric, axillary brachial and cervical lymph nodes of an adult rat, approximately $0.5–1 \times 10^8$ viable lymphocytes can be expected.

44.4. Spleen (Lymphocytes)

This is a large organ and should be cut up into three or four smaller pieces in a little BSS or PBS. Each piece is teased with forceps to liberate some of the cells, and the residual tissue is pressed gently through wire gauze (mesh No. 100; see above). The spleen cell suspension will be contaminated with rbc, and these can be removed with Tris–NH$_4$Cl (see 44.2).

An alternate means of obtaining spleen cells is to homogenise the spleen in about 2 ml BSS or PBS in a manually operated homogeniser. About 10 slow passes of the plunger should be made, taking care to produce as little frothing of the fluid as possible. The cell suspension and residual tissue is decanted into a wire gauze sieve (mesh No. 100) and the tissue pressed through gently to release the remaining cells.

Viability of the splenic lymphocytes is about 80–90%, and approximately 0.5–1 $\times$ 10^9 viable cells can be expected from an adult rat spleen.

44.5. Bone Marrow (Lymphocytes)

The rat is killed and laid on its back with its leg stretched out by means of a stout dissecting needle through its foot. Each leg is treated in turn. The skin is incised from the ankle to the hip. With a No. 23 scalpel blade, the muscles of the thigh are stripped from the femur which is then severed from the hip with scissors. The muscles are now removed from the tibia, and the epiphyseal junction between the two bones is cut. The bones do not have to be cleaned of muscle tissue too thoroughly. A small hole is made with a dissecting needle or 21G needle into the cavity of the femur and tibia through the medial condyloid surface of each bone. The tip of the opposite end of each bone is cut with scissors or bone forceps to reveal the central marrow cavity. A 21G needle connected to a 5 ml syringe filled with BSS or PBS is pushed into the marrow cavity, using each end alternatively, and the bone marrow is flushed out. It is often extruded as a long string of tissue. A similar procedure is used with the humerus bone of the forelimbs. The radius bone can also be utilised but contributes only a very small amount of marrow.

The clumps of marrow cells are dissociated by taking up the suspension in a Pasteur pipette and dispelling forcefully, the procedure being repeated four or five times. The cells are washed in BSS or PBS.

The viability of bone marrow lymphocytes is about 90%, and about 2 $\times$ 10^8 cells can be obtained from an adult rat.

44.6. Thymocytes

The gland is placed in BSS or PBS and its two lobes separated. The very tip of the broad end of each lobe is cut off and the thymocytes are squeezed out of each lobe using a pair of broad-tipped smooth straight forceps. To squeeze the lobe it is held at its narrow end and, starting here, the broad forceps are repeatedly opened and closed on the lobe moving progressively towards the broad end of the lobe. In this way the contents, often in little clumps, are ejected into the medium. The cells are dissociated by expelling several times from a Pasteur pipette. If desired, the residual pieces of thymic tissue can be pressed gently against wire mesh, mesh No. 100, to release a further small supply of cells. The cells are washed three times before being resuspended for use.

The viability of thymocytes is usually greater than 90%, and approximately 2×10^8 cells can be obtained from one thymus gland.

44.7. Cells of Other Organs and Tissues

In general, tissues other than those described above cannot be disrupted successfully by mechanical means, and the viability of the cells can be extremely low. It is best to produce single-cell suspensions of such tissues (e.g. liver, kidney, solid tumours, etc.) by enzymatic means or using chelating agents. Many different enzymes have been used for this purpose, e.g. trypsin, pronase, deoxyribonuclease, papain, collagenase, hyaluronidase and pancreatin; the disodium salt of ethylenediaminetetra-acetic acid (EDTA) has been the most successful chelating agent. Probably the most successful methods have used trypsin, sometimes in combination with EDTA, and collagenase, often combined with hyaluronidase, and these two methods will now be described.

44.7.1. Use of Trypsin

(i) Prepare a solution of 0.25% trypsin in BSS, warm to 37°C.

(ii) Cut the organ into small pieces with a scalpel and mince finely with scissors or an automatic tissue slicer.

(iii) Place the tissue in a universal container and add 15 ml of warm trypsin, shake and place in a 37°C oven for 30 min. Shake again and leave for 1–2 min to allow the tissue to settle. Remove and discard supernatant containing dead cells and debris.

(iv) Add some more trypsin and stir on a magnetic stirrer for 30 min at 37°C. Allow tissue lumps to settle for 1–2 min then remove and keep supernatant. Either centrifuge the supernatant immediately and wash once to remove the trypsin, or add foetal calf serum or rat serum to 20% concentration to neutralise the trypsin and wash later.

(v) Repeat (iv) four or five times.

(vi) Pool the supernatant cells, wash three times and resuspend in BSS or PBS. Any clumps of dead cells can be removed by passing through nylon cloth, mesh No. 100.

In some cases trypsin combined with EDTA may give a superior cell suspension. The concentration of the mixture should be 0.05 g trypsin + 0.02 g EDTA in 100 ml calcium and magnesium-free PBS. The mixture is used as for trypsin alone (EDTA in a concentration of 0.25% in calcium and magnesium-free medium can also be used alone, but its effectiveness is variable).

44.7.2. Use of Collagenase

(i) Prepare collagenase solution at a concentration of 2 mg/ml in BSS or tissue culture medium containing foetal calf serum to a concentration of 20%. (A crude collagenase preparation is quite adequate.) Prepare the solution by stirring for 1 hour at 4°C. Filter through a 0.22 μm filter if sterility of the enzyme is required. The solution should be kept at 4°C and used within 24 hours.

(ii) Place the minced tissue in a 100 ml conical flask containing three or four glass beads about 5 mm in diameter.

(iii) Add 25 ml collagenase solution and place in a gently shaking water bath at 37°C for about 1 hour.

(iv) Centrifuge at 400 times gravity for 5 min and discard the supernatant which contains dead cells and debris.

(v) Add a further 25 ml collagenase and incubate on the shaking water bath for 1-1.5 hours.

(vi) Centrifuge, discard supernatant, add BSS and filter the suspension through nylon or wire mesh to remove large pieces of tissue. Wash the cells three times and resuspend. Note that some soft tissues may require less time in contact with collagenase, and two periods of 30 min may be adequate.

A combination of 2 mg collagenase and 1 mg hyaluronidase/ml may be superior in dissociating some difficult tissues.

Treatment of tissues with collagenase (plus hyaluronidase) often produces a superior cell suspension, both in viability and in number of cells, than treatment with trypsin with or without EDTA.

Perfusion of organs with enzyme preparations is a particularly gentle way of producing viable cell suspensions. This gentle treatment is often required in some biochemical studies. The method is illustrated using the liver.

The liver is perfused using a temperature and gas controlled recirculating perfusion technique (see 41.1). The enzyme of preference is collagenase at a concentration of 50 mg/100 ml calcium-free perfusion medium. Perfusion should be continued until the rate of return of perfusion medium to the reservoir is reduced by one-third indicating that the liver is "leaking" as a result of the disrupting effect of the enzyme on the interstitial tissue of the liver. The liver is removed, chopped into small 2–3 mm pieces and placed in the perfusate. This is now shaken gently for 10 min during which the surface is gassed with the 95% air:5% carbon dioxide mixture used to gas the medium during perfusion (the gas is blown on to the surface and must not be bubbled through the perfusate). The liver suspension is first gently pressed through a large aperture wire strainer (e.g. tea strainer) then through wire mesh of apertures of 250 µm, and finally through mesh of apertures of 100 µm. The isolated hepatocytes are centrifuged at 100 times gravity for 1 min and washed three times with BSS or Krebs buffer, centrifuging at 100 times gravity, for 1 min before being resuspended each time.

45. COUNTING OF CELLS

After preparing a single cell suspension, it will often be necessary to obtain a count of the number of cells present in a unit volume. This is most conveniently done using a counting chamber (haemocytometer— see Equipment Index). There are various types of haemocytometer, but an improved Neubauer haemocytometer of the bright-line rhodinium-plated variety is adequate for most purposes (Fig. 233). The central area, consisting of 25 groups of 16 small squares, is usually used for counting rbc (but can be used to good effect to count other types of cells when they are in large numbers) and the outer four groups consisting of 16 large squares each are used for wbc and small numbers of other cell types.

It is usual to dilute blood before counting because of the large number of cells present, and rbc and wbc diluting fluids are used. For rbc the diluting fluid is a solution of 1% formalin (10 ml/l using 40% formaldehyde) in 31.3 g/l of trisodium citrate. For wbc a particular requirement is that if there is contamination by rbc then these must be lysed by the diluting fluid. An excellent fluid of this kind is Randolph's solution, which is a mixture of two solutions and which is prepared fresh and must be used within 4 hours of mixing.

(i) Solution A: methylene blue (0.1 g), propane-1,2-diol (100 ml), distilled water (100 ml).

(ii) Solution B: phloxine (0.1 g), propane-1,2-diol (100 ml), distilled water (100 ml).

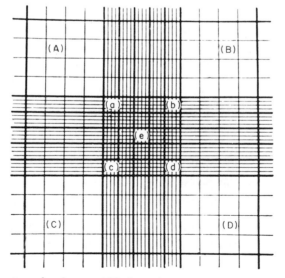

Fig. 233. Grid system of an improved Neubauer haemocytometer.

Equal volumes are mixed for use. Eosinophil granules appear red and the nuclei of wbc appear green–blue in colour. After addition of the diluting fluid the cells should be left for about 10 min to stain and settle out in the counting chamber. Another diluting fluid in common use is 2% acetic acid coloured pale violet with a little gentian violet.

For rbc counts, blood should be diluted 200 times, and this is most conveniently done using a special rbc pipette. For wbc a pipette is available which produces a dilution of 20. For counting rbc the diluted blood is placed under the coverslip of the counting chamber and the rbc in the four outer groups and the central group of the 25 groups of the central ruled area (viz. a, b, c, d and e in Fig. 233, i.e. 80 small squares) are counted, and the red cell count in millions per millilitre of blood is calculated as:

$$N \times \frac{1}{0.02} \times 200 \times 1000$$

where N is the number of cells counted in the 80 small squares of the five central groups. Each small square is 0.1 mm in depth, and 80 squares has a volume of 0.02 mm^3. The dilution factor is 200 if a rbc pipette is used. For counting cells using a wbc pipette, a minimum of 200 cells must be counted in one or more of the four outer groups of the haemocytometer (viz. A, B, C and D in Fig. 233) and the number of cells in one group (16 squares) is noted (or mean number in one group if more than one group has been used to count the minimum number of 200 cells). The number of cells per millilitre is calculated as:

$$N \times 10 \times 20 \times 1000$$

where N is the number of cells (or mean number) in one group of 16 squares. The area of such a group is 1 mm^2 and has a volume of 0.1 mm^3. The dilution factor is 20. Note that if a wbc pipette is not used then the dilution factor will vary according to the dilution made.

46. TESTING FOR CELL VIABILITY

Cells prepared as single cell suspensions will not all be viable. Since the cell membrane of living cells does not allow the entry of certain dyes, the exclusion of dyes by the cell is used frequently to discern living from dead cells which take up the dye. Trypan blue (see Equipment Index) has been the dye most often favoured and the method is as follows:

(i) Prepare a 0.1% filtered solution of trypan blue in PBS or saline.

(ii) Mix the cell suspension and the trypan blue solution in equal volumes, or use the trypan blue as the diluting fluid if a wbc pipette is employed (i.e. 20 times as much dye as cell suspension is used in the pipette).

(iii) Place in a haemocytometer within 15 min of mixing and immediately count the stained (dead) and unstained (viable) cells using a hand tally or other suitable counter. A minimum of 200 nucleated cells in total (i.e. live plus dead cells) must be counted to achieve acceptable accuracy. The percentage viability and the number of viable cells can then be calculated.

Vital dyes other than trypan blue can be used, and two such dyes are eosin Y and lissamine green, both as 0.1% solutions.

47. METHODS OF EUTHANASIA

Rats can be killed for a number of reasons, e.g. autopsy, biopsy, histology of tissues, collection of body fluids, disposal. In selecting a method, it is, therefore, important to take into account the purpose for which the animal is being killed. Whichever method of euthanasia is to be used, a prime requisite is that it must be carried out humanely, causing only the absolute minimum amount of anxiety and pain to the animal. The following methods are routinely used for the rat.

47.1. Overdose of Anaesthetic

An overdose of pentobarbitone (e.g. 120 mg/kg body weight) given intraperitoneally will kill rats after they first pass into a gentle sleep. Their necks should be dislocated subsequently to ensure that they are dead.

47.2. Carbon Dioxide

The rat should be placed in a Perspex chamber, or a metal or wooden box with a glass or Perspex lid so that it can be observed during exposure to the carbon dioxide. Carbon dioxide should be piped into the chamber from a compressed gas cylinder, so that the animal is exposed to a slowly rising concentration of the gas. This will ensure that the animal loses consciousness because of the effects of carbon dioxide on the CNS, rather than suffocating because of immediate exposure to 100% CO_2. It seems acceptable to carry out euthanasia of large numbers of animals using this technique. If large numbers of animals need to be killed, then it may be useful to build a euthanasia chamber which is large enough to accommodate the animals whilst still housed in their normal cages. This avoids some of the stress which can result from handling and transfer to a new environment.

47.3. Decapitation (Guillotine)

Guillotines are available commercially (Fig. 234—see Equipment Index) and rats must be lightly anaesthetised or stunned before they are placed in the guillotine and their heads chopped off. The heads of anaesthetised or stunned rats can also be removed using a pair of shears or stout scissors. Decapitation is usually a fairly messy procedure, and one which may be found distressing by the operator. If it is to be carried out humanely, it is important that the procedure is undertaken skilfully and rapidly. If, because of the requirements of the experimental procedure, the rats are to be killed without prior stunning or anaesthesia, it is useful to restrain the animal in a disposable polythene restrainer (Decapicone, see Equipment Index and Fig. 8).

Fig. 234. A guillotine.

47.4. Cervical Dislocation

This is a completely humane way of killing rats in spite of the potentially unpleasant psychological effect on the investigator. All rats should first be stunned and made unconscious before the neck is broken. This is because it is difficult to catch the animal first time in just the right position to break its neck. Having to make several attempts before success is unacceptable as it causes great anxiety and pain in the rat and is consequently inhumane. Apart from this, a distressed rat is also a potential danger to the investigator.

To stun a rat, the investigator should swing the rat by its tail or by the lower part of its body, so that the back of its head comes into forceful contact with a hard object, such as a table edge. The procedure must be carried out precisely so that only the head makes contact with the hard object. When attempting to gain experience in this technique, initially only deeply anaesthetised animals should be killed in this way.

After stunning the rat, a pair of stout scissors or an iron rod should be placed quickly across the neck and pressed down hard while the body is jerked backwards by pulling on the tail. This dislocates the neck. It is also sufficient to press down hard on the neck with an iron rod using both hands as this smashes the spinal column. These procedures must be carried out quickly as rats usually go into convulsions within seconds of being stunned, and thrash around to an extent which makes it less easy to locate the neck for the purposes of breaking it. Also, after stunning, rats may bleed copiously from the nose and mouth, and if allowed to go into convulsions, the blood gets distributed everywhere.

47.5. Microwaves

In certain circumstances, when it is necessary to arrest brain metabolism rapidly, microwave irradiation may be used to kill rats. This technique requires specially constructed apparatus to provide a focused source of microwaves (Schneider *et al.*, 1982). It is *not* acceptable to use a domestic microwave oven.

47.6. Choice of Method

Various histopathological changes may occur depending on the method of euthanasia used. Some of these are self-evident, e.g. stunning, cervical dislocation and decapitation will damage either the brain or the tissues of the cervical region and these methods are contraindicated if these areas are to be studied. All methods of euthanasia affect the lungs to some degree. In most cases this is simply mild congestion of the alveolar capillaries. In addition, carbon dioxide affects the ultrastructure of the liver. Intraperitoneal sodium pentobarbitone produces splenomegaly, focal congestion of the intestinal serosa and subcapsular necrosis of the liver and kidney. In general, euthanasia by overexposure to carbon dioxide or i.p. injection of sodium pentobarbitone is most suitable for pulmonary studies, whereas cervical dislocation and decapitation are more suitable for examination of the abdominal viscera and obtaining biochemically normal (unadulterated) tissues and cells.

Chapter Seven

VITAL STATISTICS AND MISCELLANEOUS INFORMATION

The establishment of normal values for physiological parameters is essential for the employment of any species as an experimental subject. However, for a laboratory species such as the rat, of which there are a large number of strains and substrains, there can be physiological variation between them which is often very marked. Also, there is normal biological variation within a particular strain or substrain and a variation may also be apparent between the sexes. For these reasons it is not possible to give accurate values for any particular parameter. However, it is very useful to have some figure to work with, particularly if the investigator is entirely ignorant of any value at all. The values given in this section are intended therefore to act only as a guide. They apply most closely to an adult rat about 3 months old and, unless specifically stated, not to any particular strain of rat. In most instances the values for male and female rats are similar, but where they vary markedly this has been indicated. An extensive summary of clinical chemistry data can be obtained from Loeb and Quimby (1989).

48. HAEMATOLOGICAL DATA

Total blood volume (ml/100 g body wt)	7
Plasma volume (ml/100 g body wt)	3.5
Haematocrit (i.e. packed cell volume; ml/100 ml)	47
Haemoglobin (g/100 ml)	15.7
Erythrocyte sedimentation rate (mm/h)	1.6 ($\male$), 0.7($\female$)
Osmolality (mmol/kg)	321
Coagulation time (min)	2.5
Clot retraction time (min to completion)	60
Red blood cells (x 10^6/mm^3)	8.9
White blood cells (x 10^3/mm^3)	9.9
Neutrophils (x 10^3/mm^3)	2.4
Lymphocytes (x 10^3/mm^3)	7.5
Monocytes (x 10^3/mm^3)	0.02
Basophils (x 10^3/mm^3)	0.02
Eosinophils (x 10^3/mm^3)	0.03
Platelets (x 10^3/mm^3)	340
pH (whole blood)	7.36
Specific gravity (whole blood)	1.05

49. CLINICAL CHEMISTRY DATA (determined for the male Wistar rat)

49.1. Plasma

Total proteins (g/l)	63
Albumin (g/l)	28
Alpha 1 globulins (g/l)	4.6
Alpha 2 globulins (g/l)	3.5
Beta globulins (g/l)	5
Gamma globulins (g/l)	4.4
Urea (mmol/l)	6.9
Ureate (mmol/l)	0.6
Glucose (mmol/l)	10.1
Creatinine (μmol/l)	42.5

Creatinine clearance (ml/min)	1.2
Total lipids (g/l)	2.3
Phospholipid (g/l)	0.05
Cholesterol (mmol/l)	1.9
Neutral fat (g/l)	0.8
Bilirubin (μmol/l)	2
Aspartate aminotransferase (i.u./l)	82
Creatine kinase (i.u./l)	368
Gamma-glutamyl transpeptidase (i.u./l)	10
Alpha-hydroxybutyrate dehydrogenase (i.u./l)	71
Alkaline phosphatase (i.u./l at 37°C)	200
Iron binding capacity (μmol/l)	101
Fe^{3+} (μmol/l)	28
Na^+ (mmol/l)	135
K^+ (mmol/l)	4.9
Ca^{2+} (mmol/l)	2.6
Cu^+ (μmol/l)	17.8
Mg^{2+} (mmol/l)	1.3
Li^+ (μmol/l)	<0.07
Cl^- (mmol/l)	100
PO_4^{3-} (mmol/l)	2.3
Colloid osmotic pressure (cm water)	26
Specific gravity (cm water)	1.21

49.2. Urine

Volume (ml/24 h)	15–30
Protein (mg/24 h)	<20
Urea (mmol/l)	442.5
Ureate (mmol/l)	1.7
Creatinine (μmol/l)	6.2
Na^+ (mmol/l)	229
K^+ (mmol/l)	149.5
Ca^{2+} (mmol/l)	0.7
PO_4^{3-} (mmol/l)	27.1
Osmolality (mmol/l)	3044
pH	6.2
Specific gravity	8

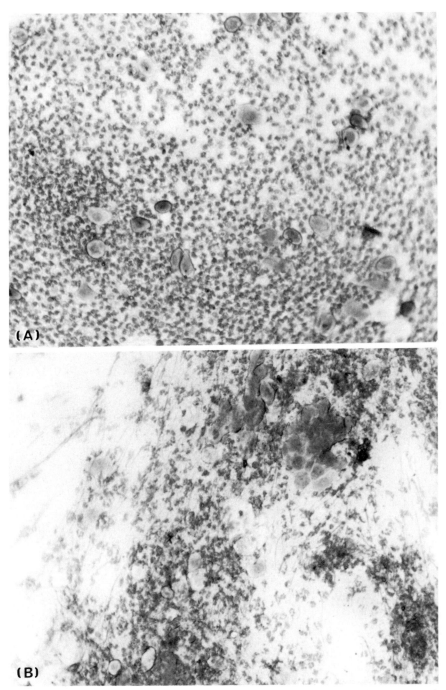

Fig. 235. Vaginal smear cell patterns during the oestrus cycle: (A) dioestrus—mostly polymorphonuclear leucocytes present with little or no mucus, (B) early pro-oestrus—polymorphonuclear leucocytes, nucleated cells and a few cornified epithelial cells present

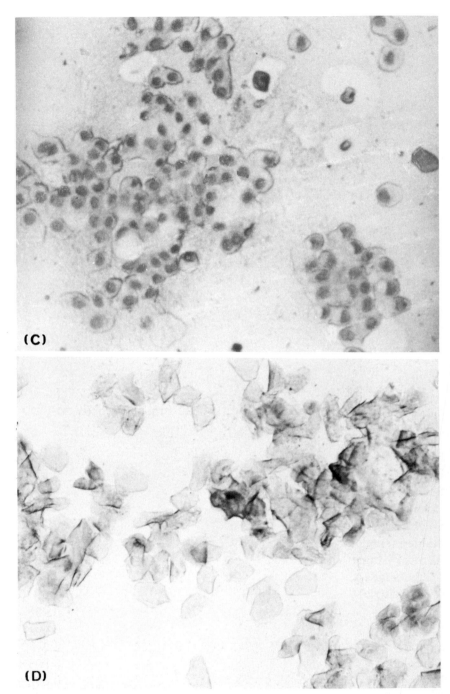

together with stringy mucus, (C) late pro-oestrus—only nucleated cells present, (D) oestrus—cornified epithelial cells only present except in the very early stages when a few irregularly shaped nucleated cells may also be seen.

50. REPRODUCTIVE PARAMETERS

Polyoestrus female sex cycle type. Stages of cycle and approximate duration:

Stage 1 dioestrus	6 h
Stage 2 pro-oestrus (early)	60 h
Stage 3 pro-oestrus (late)	12 h
Stage 4 oestrus	10–20 h
Stage 5 metoestrus	8 h
Total duration of a single cycle	4–5 days
Vaginal smear patterns during the cycle	See Fig. 235
Age at puberty (and first mating)	6–11 weeks
Gestation period	21–23 days
Length of pseudopregnancy	12 days
Number of young/litter	6–10
Age at weaning	21 days

51. ORGAN WEIGHTS (wet weight in g/100 g body wt)

Heart	0.4
Lungs	0.6
Brain	1
Liver	3
Spleen	0.2
Kidney (single)	0.4
Adrenal (single)	0.02
Testis (single)	0.5
Ovary (single)	0.05
Pituitary	0.005
Pineal	0.001
Stomach and intestines	2.3
Thyroid	0.005
Lymph node (popliteal)	0.002
Thymus	0.07
Prostate	0.16
Seminal vesicles (semen removed)	0.3

52. MISCELLANEOUS DATA

Body temperature (°C)	38.1
Heart rate (beats/min)	328
Respiration rate (breaths/min)	97
Systolic blood pressure (mmHg)	116–180
Diastolic blood pressure (mmHg)	90
Lethal X- or γ-irradiation dose (Grays)	10
pH of gastric secretion (6 h collection)	1.3–2.6
pH of gastric contents (collected by lavage)	2.5–4
Chromosome number	42
Life span (laboratory conditions; years)	2–4
Daily food intake (g/100 g body wt)	5
Daily water intake (ml/100 g body wt)	10
Rate of growth	See Fig. 236

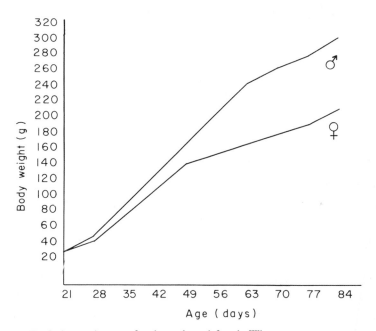

Fig. 236. Typical growth curve for the male and female Wistar rat.

53. METHODS OF IDENTIFICATION

53.1. Ear marking

A useful means of marking rats permanently for identification is to punch a hole and/or cut a wedge in the ears. The rat is then identified by the number and/or position of these and by the ears used. A system using wedges alone which can be cut into the ears with a pair of small sharp scissors is shown in the following table:

	Number of wedges cut			Total number of
Rat	Left ear	Right ear	Identification	wedges cut
1	0	0	P (plain)	0
2	1	0	L_1	1
3	0	1	R_1	1
4	1	1	B_1 (both)	2
5	2	0	L_2	2
6	0	2	R_2	2
7	2	1	L_2R_1	3
8	1	2	R_2L_1	3
9	3	0	L_3	3
10	0	3	R_3	3
11	2	2	B_2	4
12	3	1	L_3R_1	4
13	1	3	R_3L_1	4
14	4	0	L_4	4
15	0	4	R_4	4

For marking more than 15 rats a system incorporating punch holes (produced by a special ear punch—see Equipment Index) combined if desired with wedges, placed variously about the ears can be used. This is shown in Fig. 237.

53.2. Dyes

Rats may be marked for temporary identification by using commercially available waterproof marker pens. A suitable code when using these on the tail is shown in Fig. 238.

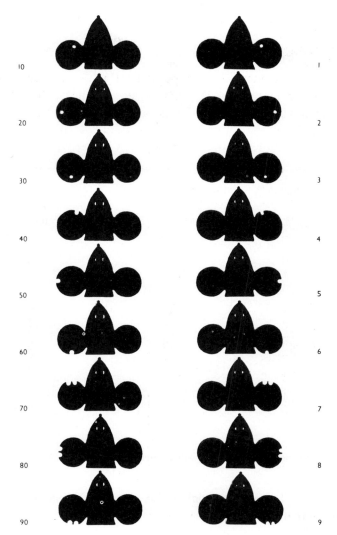

Fig. 237. Code for marking the ears of rats with holes and wedges for identification.

53.3. Ear Tags

Rats can be individually identified using metal ear tags (see Equipment Index). These numbered tags are suitable for short-term (up to 14 days) identification. Unfortunately, a significant number of animals will remove

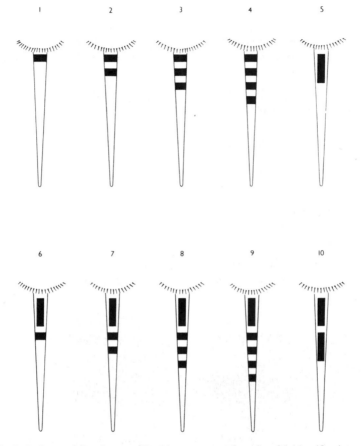

Fig. 238. Code for marking the rat tail with a waterproof marker for identification.

their tags. Since removal invariably also damages the ear, this technique is not recommended.

53.4. Electronic Tagging

Rats can be permanently and individually identified by subcutaneous implantation of a microchip device (see Equipment Index). Each device has a unique identification number which is read using a special detector. The technique is simple and reliable, but its relatively high cost limits widespread use.

53.5. Tattoo

The procedure for tattooing numbers on the tail using a commercially available tattooing machine (Fig. 239—see Equipment Index) is particularly useful if large numbers of animals are involved or animals are to be maintained for a prolonged period. These numbers make identification particularly easy and quick. However, experience and practice are required to use the tattooing machine so that clear numbers are produced. To carry out the procedure the manufacturer's instructions should be followed.

53.6. Neonatal rats

For marking neonatal rats up to about 10 days of age, it has been usual to remove one or more digits from one or both feet using sharp scissors and a simple code. The procedure is probably painful but a suitable alternative for coloured strains of rats is not readily available. However,

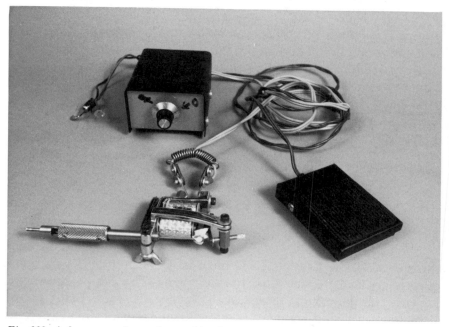

Fig. 239. A foot-operated tattooing machine for use with rats.

for neonatal albino rats, this procedure is unnecessary as they can be successfully tattooed using a simple code involving placing appropriately one or more dots about the body. When handling neonates it is necessary to take steps to prevent the mother subsequently cannibalising her young (see 2.15.7.)

54. DETERMINATION OF THE STAGE OF THE OESTROUS CYCLE

A small metal spatula, blunted with rounded edges, is dipped into water and placed deep into the vaginal opening. The walls of the vagina are scraped gently with a scooping motion of the spatula and the moist scrapings collected by the spatula are spread on to a clean glass slide and allowed to air dry (N.B. overzealous scraping should be avoided). As an alternative, a small cotton wool bud on a stick or held in forceps can be moistened with water and rotated within the vagina and then wiped on a slide, or a drop of water pipetted in and out of the vagina two or three times will collect cells which can be transferred to a slide and allowed to dry. The slide is then dipped into a filtered 0.1% (w/v) aqueous solution of methylene blue, the surplus of which is carefully removed from the slide under running tap water. The slide is up-ended, allowed to dry and then examined under a low powered microscope. The stage of the cycle is gauged from the type(s) of cell(s) found (see Fig. 235).

55. DETERMINATION OF PREGNANCY

After mating, the female rat will often be found to have a hard white plug of ejaculated semen lodged in the vagina. This can be determined by either inspecting the female, or by finding the vaginal plug on the cage tray (in a grid-bottomed cage) when it becomes dislodged some time after mating. In some cases, a vaginal plug fails to form or is lost and a smear of the vaginal contents must be taken and examined microscopically (see 54 but omit staining with methylene blue). The finding of spermatozoa indicates that mating has taken place. The examination must be thorough as sometimes only one or two spermatozoa can be found.

If there is no requirement for timed mating, then males and females should be housed together continuously or put together in the evening and vaginal smears taken or a plug looked for early the following morning. Copulation usually takes place during the dark period. If precise timed mating is required, the following procedure is very accurate and up to 95% of females should become pregnant.

Male and female rats are kept in separate grid-bottomed cages in the same room for at least 7–10 days. During this period they are acclimatised to a light cycle of lights off between 02:00 hours and 12:00 hours and lights on between 12:00 hours and 02:00 hours. Following acclimatisation and while maintaining this light cycle vaginal smears are taken daily at between 08:00 and 09:00 hours, each rat being exposed to light for no more than 30 seconds while this is being done. Females showing an oestrous smear are then placed with a male during the final 2 hours of the dark period, between 10:00 and 12:00 hours. The females are then removed and inspected for a vaginal plug. The cage tray is also inspected for a vaginal plug which may have dropped into it and this is made simpler if the tray is lined with brown paper. If no vaginal plug is found it is assumed that successful copulation has not taken place and the female is returned to the male only at the next oestrous, if required. If copulation is successful, pregnancy is assumed to have been initiated at 11:00 hours ± 1 hour.

56. PREPARATION OF A SMOOTH DIET FOR TUBE FEEDING

A calorifically adequate diet for the rat which can be suspended in water and fed through a small polythene tube (or which can be fed as a dry powdered diet), is composed of the following ingredients:

	g/kg of diet
Casein—vitamin free (finely ground to pass through a No. 100 mesh sieve)	200
L-Cystine	3
Sucrose	250
Olive or corn oil	50
Salt mixture*	30
Choline hydrochloride	6

	mg/kg of diet
Thiamine	10
Riboflavine	16
Nicotinic acid	100
Pyridoxine	10
Calcium pantothenate	20
Inositol	200
p-Aminobenzoic acid	100
Folic acid	2
Menaphthone	10
Vitamin B_{12}	0.1
Vitamin E	200
Vitamin A + D (given as Rovimix, 10 mg/kg)†	
Vitamin A (i.u.)	500
Vitamin D (i.u.)	325

Another method for preparing a smooth uniform diet suspension is as follows: 40 g of powdered diet made by pulverising normal rat pelleted diet is added to 50 ml of water and mixed vigorously for about 1 min. A portion of this mixture (e.g. 40 ml) is placed in a 50 ml syringe which is placed vertically in a tissue press. A second 50 ml syringe, to which is attached a large bore needle (e.g. 16G), is placed immediately below the nozzle of the first syringe to catch the effluent. The plunger of the first syringe is slowly depressed by rotating the lever of the tissue press, easily extruding the diet into the barrel of the second syringe. This process breaks up all the large lumps. The procedure is

*Glaxo Laboratories Ltd., Greenford, Middlesex, UK.
†Roche Products Ltd., Welwyn Garden City, Hertfordshire, UK.

now repeated with the second syringe to produce a finely divided diet which is collected in a suitable container.

The liquid diet from either the first or the second method has a calorific density of about 1.9 kcal/ml and an adequate calorific intake will be provided by feeding adult rats an amount not exceeding the capacity of their stomachs, three or four times a day.

57. DOSE RATES OF HORMONES IN THE RAT

Hormone	Dose rate	Comments
Aldosterone	0.5 mg total dose/day, s.c.	Following adrenalectomy to return plasma Na to normal
Diethylstilboestrol	1 mg total dose/day s.c.	To stimulate ovarian growth
	1 μg total dose/day s.c.	To induce vaginal cornification after ovariectomy
Hydrocortisone acetate	0.1 mg total dose twice daily, s.c.	Replacement therapy after adrenalectomy
	15 mg total dose/day s.c.	To correct hypoglycaemia following adrenalectomy
Insulin	6–12 units twice daily	For maintenance therapy after pancreatectomy
Oestradiol benzoate	15 μg/kg/day s.c.	Replacement therapy after ovariectomy
Oxytocin	5 i.u. total dose s.c.	To stimulate milk ejection in lactating rat

| Progesterone | 20 mg total dose/day s.c. | To maintain pregnancy after ovariectomy |
| Testosterone | 50 mg/kg/day s.c. | Replacement therapy after castration |

58. DOSE RATE OF ANTIBIOTIC, ANTIBACTERIAL AND ANTIPARASITIC AGENTS IN THE RAT

	Dose rate
Antibiotic and antibacterial	
Ampicillin	150 mg/kg s.c. twice daily
Amoxycillin	150 mg/kg i.m. once daily
Benzylpenicillin	20 000 units/kg p.o. once daily
Cephalexin	60 mg/kg p.o., 15 mg/kg s.c. once daily
Chloramphenicol	20–50 mg/kg p.o. twice daily
	10 mg/kg i.m. twice daily
Co-trimazine	0.5 ml/kg s.c. once daily
Gentamicin	4.4 mg/kg i.m. twice daily
Griseofulvin	25 mg/kg p.o.
Neomycin	2 g/l drinking water
Oxytetracycline	800 mg/l drinking water
Sulphadimidine	200 mg/l drinking water
Tetracycline	15–20 mg/kg p.o. twice daily
Tylosin	10 mg/kg i.m. once daily
Endoparasiticides	
Dimetridazole	1 g/l drinking water
Ivermectin	200 μg/kg s.c.
Niclosamide	100 mg/kg p.o.
Piperazine	2 g/l drinking water for 7 days
Thiabendazole	200 mg/kg p.o. for 5 days
Ectoparasiticides + other agents	
Bromocyclen	0.2–0.26% wash

59. SOLUTIONS

Alsever solution (g/l)

Sodium chloride	4.2
Trisodium citrate (dihydrate)	8.0
Glucose	20.5

Adjust to pH 6.1 with 10% citric acid; for use add in the proportion 1.2 ml to 1 ml blood.

Acid–citrate–dextrose solution (g/l)

Trisodium citrate (dihydrate)	22
Citric acid	8
Glucose	24.5

For use add in the proportion 0.15 ml to 1 ml blood.

Hanks balanced salt solution pH 7.2 (g/l)

Sodium chloride	8
Potassium chloride	0.4
Magnesium sulphate ($7H_2O$)	0.1
Disodium hydrogen phosphate	0.048
Calcium chloride ($2H_2O$)	0.185
Sodium bicarbonate	0.35
Potassium dihydrogen phosphate	0.06
Magnesium chloride ($6H_2O$)	0.1
Glucose	1
Phenol red (if required)	0.01

Krebs buffer pH 7.4 (g/l)

1.	Sodium chloride	9
2.	Potassium chloride	11.5
3.	Calcium chloride	12.2
4.	Potassium dihydrogen phosphate	21.1
5.	Magnesium sulphate ($7H_20$)	38.2
6.	Sodium bicarbonate	13

For use mix 100 parts solution 1, four parts solution 2, three parts solution 3, one part solution 4, one part solution 5, and 21 parts solution 6. Gas with 95% oxygen:5% carbon dioxide for 20–30 min if required.

Phosphate buffered saline pH 7.4 (Dulbecco) (g/l)

Sodium chloride	8
Potassium chloride	0.2
Disodium hydrogen phosphate (anhydrous)	1.15
Potassium dihydrogen phosphate (anhydrous)	0.2
Calcium chloride	0.0005
Magnesium chloride	0.0005

Omit calcium chloride and magnesium chloride if a Ca^{2+}- and Mg^{2+}-free solution is required.

Ringer solution (g/l)

Sodium chloride	9
Potassium chloride	0.42
Calcium chloride	0.25

Tris-ammonium chloride pH 7.2 (g/l)

Tris-(hydroxymethyl)-methylamine	20.6
Ammonium chloride	8.3

Adjust to pH 7.2 with 0.1 N hydrochloric acid; for use add 10 ml Tris to 90 ml ammonium chloride. Add approximately 10 parts of this mixture to one part of blood and leave at room temperature for 10 min to lyse rbc before centrifuging and washing pellet three times.

REFERENCES

Abildgaard, C. F. (1964). A simple apparatus for holding rats. *Lab. Anim. Care* **14**, 235.

Alexander, J. I. and Hill, R. G. (1987). "Postoperative Pain Control." Blackwell Scientific Publications, Oxford.

Anderson, R. M. and Romfh, R. F. (1980). "Technique in the use of surgical tools". Appleton-Century-Crofts, New York.

Animal Welfare Act (1966) (amended 1970, 1976 and 1985). Federal Register, Vol. 32. Obtained from United States Department of Agriculture, Room 565, Federal Building, 6505 Belcrest Road, Hyattsville, Maryland 20782, USA.

Animals (Scientific Procedures) Act, 1986 (1986). HMSO, London.

Archer, R. K. and Riley, J. (1981). Standardised method for bleeding rats. *Lab. Anim.* **15**, 25.

Association of Veterinary Teachers and Research Workers (AVTRW) (1989). Guidelines for the recognition and assessment of pain in animals. University Federation for Animal Welfare, Potters Bar, Herts.

Bartoszyk, G. D. and Wild, A. (1989). B-vitamins potentiate the antinociceptive effect of dislofenac in carrageenin-induced hyperalgesia in the rat tail pressure test. *Neurosci. Lett.* **101**, 95.

Beynen, A. C., Baumans, V., Bertens, A. P. M. G., Havenaar, R., Hesp, A. P. M. and Van Zutphen, L. F. M. (1986). Assessment of discomfort in gallstone-bearing mice: a practical example of the problems encountered in an attempt to recognise discomfort in laboratory animals. *Lab. Anim.* **21**, 35.

Bhatt, P. N., Jacoby, R. O., Morse, H. C. and New, A. E. (eds) (1986). "Viral and Mycoplasmal Infections of Laboratory Rodents." Academic Press Inc., Orlando.

Birnbaum, D. and Hall, T. (1961). An electro-ejaculation technique for rats. *Anat. Rec.* **140**, 49.

Blowers, R. (1964). Hospital infection. *In* "Clinical Surgery I. General Principles and Breast" (Rob, C. and Smith, R. eds). Butterworths, London.

Boetzner, P., Fehm, H. L., Voight, K. H., Schleyer, M. and Pfuffer,

E. F. (1972). Measurement of biological activity in various growth hormone preparations using a modified tibia test. *Acta Endocr., Copenh.* **70**, 231.

Born, C.T. and Moller, M. L. (1974). A simple procedure for long-term intravenous infusion in the rat. *Lab. Anim. Sci.* **24**, 355.

Brakkee, J. H., Wiegand, V. M. and Gispen, W. H. (1979). A simple technique for rapid implantation of a permanent cannula into the rat brain ventricular. *Lab. Anim. Sci.* **2**, 78.

Buckley, P. F. (1985). Somatic nerve blocks for postoperative analgesia. *In* "Acute Pain" (Smith, G. and Covino, B. G., eds), pp. 205–227. Butterworths, London.

Buelke-Sam, J., Holson, J. F., Bazare, J. J. and Young, J. F. (1978). Comparative stability of physiological parameters during sustained anaesthesia in rats. *Lab. Anim. Sci.* **28**(2), 157.

Burns, J. and Robbins, B. G. (1972). A modified silver clip used in the induction of renal hypertension in the rat. *J. Pharmacol. Pharmacol.* **24**, 86.

Campbell, P. N. and Sargent, J. R. (1969). *In* "Techniques in Protein Biosynthesis" Vol. 2, Ch. 5. Academic Press, London and New York.

Caulfield, M. P., Clover, K. F., Powers, D. A. and Savage, T. (1983). A rapid and convenient freehand method for the implantation of cerebroventricular cannulae in rats. *J. Pharmacol. Meth.* **9**, 231.

Chapman, R. C. (1985). Psychological factors in postoperative pain. *In* "Acute Pain" (Smith, G. and Covino, B. G., eds), p. 22. Butterworths, London.

Chiou, W. L. (1989). The phenomenon and rationale of marked dependence of drug concentration on blood sampling site. *Clin. Pharmacokinet.* **17**, 175.

Chipman, J. K. and Cropper, M. C. (1977). A technique for chronic intermittent bile collection from the rat. *Res. Vet. Sci.* **22**, 366.

Cooper, J. E. (1982) Dealing with non-domesticated species. *In* "A Manual of Practice Improvement" (Fry, P. D., ed.). BSAVA, London.

Costa, D. L., Lehmann, J. R., Harold, W. M. and Drew, R. T. (1986). Transoral tracheal intubation of rodents using a fibreoptic laryngoscope. *Lab. Anim. Sci.* **36**, 256.

Cowan, A., Lewis, J. W. and McFarlane, I. R. (1977). Agonist and antagonist properties of buprenorphine, a new antinociceptive agent. *Br. J. Pharmacol.* **60**, 537.

Cunliffe-Beamer, T.L. (1990). Surgical Techniques. *In* "Guidelines for the Well-being of Rodents in Research" (Guttman H. N., ed). Scientists Center for Animal Welfare, Bethesda.

Dalton, R. G., Touraine, J. L. and Wilson, T. R. (1969). A simple technique for continuous intravenous infusion in rats. *J. Lab. Clin. Med.* **74**, 813.

Desjardins, C. (1986). Indwelling vascular cannulas for remote blood sampling, infusion, and long-term instrumentations of small laboratory animals. *In* "Methods of Animal Experimentation", Vol. VII, Part A (Gay, W. I. and Heavner, J. E., eds), p. 143. Academic Press, Orlando.

DeWildt, D. J., Hillen, F. C., Rauws, A. G. and Sangster, B. (1983). Etomidate-anaesthesia, with and without fentanyl, compared with urethane-anaesthesia in the rat. *Br. J. Pharmacol.* **79**, 461.

Donnelly, T. H. and Stark, D. M. (1985) Susceptibility of laboratory rats, hamsters and mice to wound infection with *Staphylococcus aureus*. *Am. J. Vet. Res.* **46**, 2634.

Dum, J. E. and Herz, A. (1981) In vivo receptor binding of the opiate partial agonist, buprenorphine, correlated with its agonistic and antagonistic actions. *Br. J. Pharmacol.* **74**, 627.

Eger, E. I. (1981). Isoflurane: a review. *Anaesthesiology* **55**, 559.

Ellis, G. B. and Desjardins, C. (1982). Male rats secrete luteinizing hormone and testosterone episodically. *Endocrinology* **110**, 1618.

Fagin, K. D., Shinsako, J. and Dallman, M. F. (1983). Effects of housing and chronic cannulation on plasma ACTH and corticosterone in the rat. *Am. J. Physiol.* **245**, E515.

Fairney, A. and Weir, A. A. (1970). The effect of abnormal maternal plasma calcium levels on the offspring of rats. *J. Endocrinol.* **48**, 337.

Farris, H. E. and Snow, W. S. (1987). Comparison of isoflurane, halothane or methoxyflurane with combinations of ketamine HCl and acepromazine for anaesthesia of the rat. *Lab. Anim. Sci.* **37**(4), 520.

Field, K. J. and Lang, C. M. (1988). Hazards of urethane (ethyl carbamate): a review of the literature. *Lab. Anim.* **22**, 255.

Flecknell, P. A. (1984). The relief of pain in laboratory animals. *Lab. Anim.* **18**, 147.

Flecknell, P. A. and Liles, J. H. (1991). The effects of surgical procedures, halothane anaesthesia and nalbuphine on locomotor activity and food and water consumption in rats. *Lab. Anim.* **25**, 50.

Flecknell, P. A. and Liles, J. H. (1992). The evaluation of locomotor activity and food and water consumption as a method of assessing post-operative pain in rodents. *In* "Animal Pain" (Short, C. E. and Van Poznak, A., eds). Churchill Livingstone, New York.

Flecknell, P. A. and Mitchell, M. (1984). Midazolam and fentanyl-fluanizone: assessment of anaesthetic effects in laboratory rodents and rabbits. *Lab. Anim.* **18**, 143.

Flecknell, P. A., Liles, J. H. and Wootton, R. (1989). Reversal of fentanyl/fluanisone neuroleptanalgesia in the rabbit using mixed agonist/antagonist opioids. *Lab. Anim.* **23**, 147.

Fleischman, R. W., McCracken, D. and Forbes, W. (1977). Adynamic ileus in the rat induced by chloral hydrate. *Lab. Anim. Sci.* **27** (2), 238.

Fox, M. W. (1986). "Laboratory Animal Husbandry." State University of New York Press, Albany.

Garner, D., McGivern, R., Jagels, G. and Laks, M. M. (1988). A new method for direct measurement of systolic and diastolic pressures in conscious rats using vascular-access-ports. *Lab. Anim. Sci.* **38**, 205.

Gay, V. L. (1965). "Methods of Animal Experimentation", Vol. 2. Academic Press, London.

Gay, V. L. (1967). A steriotaxic approach to transauricular hypophysectomy in the rat. *Endocrinology* **81**, 1177.

Gay, W. I. and Heavner, J. E. (eds) (1986). "Methods of Animal Experimentation", Vol. VII, Part A. Academic Press, Orlando.

Gellai, M. and Valtin, H. (1979). Chronic vascular constrictions and measurement of renal function in conscious rats. *Kidney Int.* **15**, 419.

Glen, J. B. (1980). Animal studies of the anaesthetic activity of ICI 35 868. *Br. J. Anaesth.* **52**, 731.

Goldschmidt, F. (1972). Reproducible topical staphylococcal infection in rats. *Appl. Microbiol.* **23**, 130.

Green, C. J. (1987). "Animal Anaesthesia." Laboratory Animal Handbooks Ltd, London.

Green, C. and Simpkin, S. (1988). "Basic Microsurgical Techniques." Royal Society of Medicine Services Ltd, 1 Wimpole Street, London W1M 8AE, UK.

Green, C. J., Halsey, M. J., Precious, S. and Wardley-Smith, B. (1978). Alphaxalone-alphadolone anaesthesia in laboratory animals. *Lab. Anim. Sci.* **12**, 85.

Griffith, Jr, J. Q. and Farris, E. J. (eds) (1942). "The Rat in Laboratory Investigation", Ch. 9. J. B. Lippincott, Philadelphia.

Guedel, A. E. (1937). "Inhalation Anaesthesia; A Fundamental Guide." MacMillan, New York.

Gupta, B. N. (1973). Technique for collecting blood from neonatal rats. *Lab. Anim. Sci.* **23**, 559.

Hall, G. M. (1985). The anaesthetic modification of the endocrine and metabolic response to surgery. *Annals of the Royal College of Surgeons of England.* Vol. 67, 25.

Halladay, S. C. (1973). An inexpensive metabolism cage for small animals. *Bull. Environ. Contam. Toxicol.* **10**, 156.

Hanwell, A. and Linzell, J. L. (1972). Validation of the thermodilution technique for estimation of cardiac output in the rat. *Comp. Biochem. Physiol.* **41**, 647.

Hard, G. C. (1975). Thymectomy in the neonatal rat. *Laboratory Animal,* **9**, 105.

Hoover-Plow, J. L. and Clifford, A. J. (1978). The effect of surgical trauma on muscle protein turnover in rats. *Biochem. J.* **176**, 137.

Hsieh, K. S. and Ota, M. (1969). Improved procedure of pinealectomy in rat. *Endocrinol. Jap.* **16**, 477.

Hu, C., Flecknell, P. A. and Liles, J. H. (1992). Fentanyl and medetomidine anaesthesia in the rat and its reversal using atipamazole and either nalbuphine or butorphanol. *Lab. Anim.* **26**, 15–22.

Hull, C. J. (1985). Opioid infusions for the management of postoperative pain. *In* "Acute Pain" (Smith, G. and Covino, B. G., eds), p. 155. Butterworths, London.

Hurov, L. (1978). "Handbook of Veterinary Surgical Instruments and Glossary of Surgical Terms." W. B. Saunders, Philadelphia.

Hurtubise, M. R., Bottino, J. C., Lawson, M. and McCreckie, K. B. (1980). Restoring patency of occluded central venous catheters. *Arch. Surg.* **115**, 212.

Ingle, D. J. and Griffith, J. Q. (1942). Surgery of the rat. *In* "The Rat in Laboratory Investigation" (Griffith Jr, J. Q. and Farris, E. J., eds), J. B. Lippincott, Philadelphia.

Jenkins, W. L. (1987). Pharmacologic aspects of analgesic drugs in animals: an overview. *J. Am. Vet. Med. Assoc.* **10**, 1231.

Johnston, B. A., Eisen, H. and Fry, D. (1991). An evaluation of several adjuvant emulsion regimens for the production of polyclonal antisera in rabbits. *Lab. Anim. Sci.* **41**, 15.

Kay, B. (1987). Opioids by infusion. *In* "Update in Opiates." *Clinical Anaesthesiology* **1**, 935.

Kehlet, H. (1978). Influence of epidural anaesthesia on the endocrine-metabolic response to surgery. *Acta Anaesthesiol. Scand. (Suppl.)* **70**, 39.

Kennedy, Jr. G. L. (1989). Inhalation toxicology. *In* "Principles and Methods of Toxicology", 2nd edn (Hayes, A. W., ed.), Raven Press, New York.

Kitchell, R. L., Erikson, H. H. and Karstens, E. (1983). "Animal Pain." American Physiological Society, Bethesda.

Knecht, C. D., Allen, A. R., Williams, D. J. and Johnson, J. H. (1987). "Fundamental Techniques in Veterinary Surgery". Churchill Livingstone, New York.

Kowalewski, K. and Chmura, G. (1969). A method permitting prolonged and repeated studies of Rat's gastric secretion. *Arch. Int. Physiol. Biochem.* **77**, 10.

Kraus, A. L. (1980). Research methodology. *In* "The Laboratory Rat" (Baker, H. J., Lindsey, J. R. and Weisbroth, S. H., eds), Vol. II. Academic Press, New York.

Krizek, T. J. and Robson, M. C. (1975). Biology of surgical infection. *Surg. Clin. North Am.* **55**, 1261.

Laboratory Animal Science Association (LASA) (1990). The assessment and control of the severity of scientific procedures on laboratory animals. Report of the Laboratory Animal Science Association Working Party, Vol. 24, No. 2, p. 97.

Lambert, R. (1965). "Surgery of the Digestive System in the Rat." Charles C. Thomas, Illinois.

Landi, M. S., Kreider, J. W., Lang, M. and Bullock, L. P. (1982). Effects of shipping on the immune function in mice. *Am. J. Vet. Res.* **43**, 1654.

Lane, D. R. (ed.) (1989). "Jones Animal Nursing", 5th edn. Pergamon Press, Oxford.

Lang, C. M. (1982). "Animal Physiologic Surgery", 2nd edn. Springer-Verlag, New York.

Leenan, F. H. H. and De Jong, W. (1971). A solid silver clip for induction of predictable levels of renal hypertension in the rat. *J. Appl. Physiol.* **31**, 142.

Leese, T., Husken, P. A. and Morton, D. B. (1988). Buprenorphine analgesia in a rat model of acute pancreatitis. *Surg. Res. Commun.* **3**, 53.

Leslie, G. B. and Conybeare, G. (1988). An improved oral dosing technique for rats. *In* "Laboratory Animal and Health for All." Proceedings of the IX ICLAS International Symposium in Laboratory Animal Science, Bangkok. (Erichsen, S., Coates, M. E. and Chatikavaniz, P. eds), The National Laboratory Animal Centre, Mahidol University, Salaya, Nakhon Pathom 73170, Thailand.

Liles, J. H. and Flecknell, P. A. (1992). The effects of buprenorphine, nalbuphine and butorphanol alone or following halothane anaesthesia on food and water consumption and locomotor movement in rats. *Laboratory Animals* (in press).

Linde, H. W. and Berman, M. L. (1971). Nonspecific stimulation of drug-metabolizing enzymes by inhalation anesthetic agents. *Anesth. Analg.* **50(4)**, 656.

Loeb, W. F. and Quimby, F. W. (1989). "The Clinical Chemistry of Laboratory Animals." Pergamon Press, New York.

Lovell, D. P. (1986). Variation in pentobarbitone sleeping time in mice. 1. Strain and sex differences. *Lab. Anim.* **20**, 85.

Lu, F. C. (1985). "Basic Toxicology." Hemisphere Publishing Corporation, Washington.

Lumb, W. V. and Jones, W. E. (1984). "Veterinary Anaesthesia." Lea and Febiger, Philadelphia.

McGee, M. A. and Maronpot, R. R. (1979). Hardarian gland dacryoadenitis in rats resulting from orbital bleeding. *Lab. Anim. Sci.* **29**, 639.

McKenzie, J. E., Anselmo, D. M. and Muldoon, S. M. (1985) Nalbuphine's reversal of hypovolaemic shock in the anaesthetised rat. *Circulation Shock* **17**, 21.

Meindl, J. D., Ford, A. J., Taperell, R. E., Bazuin, B.J., Abadi, K., Bowman, L., Smith, M. J., Dorman, H. G., Schmitt, J., Harame, D., Bousse, L., Fotowat-Ahmady, A., Shapiro, F., Gross, S. J., Midkiff, N., Akram, M. F., Gonzalez, C., Prisbe, M., Mustafa, J., Koepnick, J. H., Andrade, R. M., Claude, D. H. and Marcus, R. (1986). Implantable telemetry. *In* "Methods of Animal Experimentation", Vol. VII, Part A (Gay, W. I. and Heavner, J. E., eds). Academic Press, Orlando.

Minasian, H. (1980). A simple tourniquet to aid mouse tail venipuncture. *Lab. Anim.* **14**, 205.

Morton, D. B. and Griffiths, P. H. M. (1985). Guidelines on the recognition of pain, distress and discomfort in experimental and an hypothesis for assessment. *Vet. Rec.* **116**, 431.

Murray, W. J. and Fleming, P. J. (1972). Defluorination of methoxyflurane during anesthesia: comparison of man with other species. *Anesthesiology* **37**, 620.

Nerenberg, S. T. and Zedler, P. (1975). Sequential blood samples from the tail vein of rats and mice obtained with modified Leibig condenser jackets and vacuum. *J. Lab. Clin. Med.* **85**, 523.

Nevalainen, T., Phyhala, L., Voipio, H. M. and Virtanen, R. (1989). Evaluation of anaesthetic potency of medetomidine-ketamine combination in rats, guinea-pigs and rabbits. *Acta Vet. Scand.* **85**, 139.

Nicol, T., Vernon-Roberts, B. and Quantock, D. C. (1965). Protective effect of oestrogens against the toxic decomposition products of tribromoethanol. *Nature* **208**(**5015**), 1099.

Norris, M. L., and Turner, W. D. (1983). An evaluation of tribromoethanol (TBE) as an anaesthetic agent in the Mongolian gerbil (*Meriones unguiculatus*). *Lab. Anim.* **17**, 324.

Ogino, K., Hobara, T., Kobayashi, H. and Iwamoto, S. (1990). Gastric mucosal injury induced by chloral hydrate. *Toxicol. Lett.* **52**, 129.

O'Hair, K. C., Dodd, K. T., Phillips, Y. Y. and Beattie, R. J. (1988). Cardiopulmonary effects of nalbuphine hydrochloride and butorphanol tartrate in sheep. *Lab. Anim. Sci.* **38**, 58.

Olson, M. E. and Renchko, P. (1988). Azaperone and Azaperone-Ketamine as a neuroleptic sedative and anesthetic in rats and mice. *Lab. Anim. Sci.* **38**(3), 299.

Pakes, S. P., Lu, Y.-S. and Meunier, P. C. (1984). Factors that complicate animal research. *In* "Laboratory Animal Medicine" (Fox, J. G., Cohen, B. J. and Loew, F. M., eds). Academic Press, Orlando.

Palm, V., Boemke, W., Bayerl, D., Schnoy, N., Juhr, N.-C. and Reinhardt, H. W. (1991). Prevention of catheter-related infections by a new catheter-restricted antibiotic filling technique. *Lab. Anim.* **25**, 142.

Paré, W. P., Isom, K. E., Vincent, Jr, G. P. and Glavin, G. B. (1977). Preparation of a chronic gastric fistula in the rat. *Lab. Anim. Sci.* **27**, 244.

Pellegrino, L. J. and Cushmann, A. J. (1967). "A Steriotaxic Atlas of the Rat Brain." Appleton-Century-Crofts, New York.

Phalen, R. F. (1984). "Inhalation Studies, Foundations and Techniques." CRC Press Inc., Cleveland.

Pircio, A. W., Glys, J. A., Cavanagh, R. L., Buyniski, J. P. and Bierwagen, M. E. (1976) The pharmacology of butorphanol, a 3,14-dihydroxymorphinan narcotic antagonist analgesic. *Arch. Int. Pharmacodyn.* **220**, 231.

Popovic, V., Kent, K. M. and Popovic, P. (1963). Technique of permanent cannulation of the right ventricle in rats and ground squirrels. *Proc. Soc. Exp. Biol. Med.* **113**, 599.

Popp, M. B. and Brennan, M. F. (1981). Long-term vascular access in the rat: importance of asepsis. *Am. J. Physiol.* **241**, H606.

Riley, V. (1960). Adaptation of orbital bleeding technique to rapid serial blood studies. *Proc. Soc. Exp. Biol. Med.* **104**, 751.

Ross, B. D. (1972). "Perfusion Techniques in Biochemistry." Clarendon

Press, Oxford.

Ryer, F. H. and Walker, D. W. (1971). An anal cup for rats in metabolic studies involving radioactive materials. *Lab. Anim. Sci.* **21**, 942.

Salo, M. (1988). The relevance of metabolic and endocrine responses to anesthesia and surgery. *Acta Anaesthesiol. Belg.* **30**(3), 133.

Scharschmidt, B. and Berk, P. D. (1973). A simple device to facilitate rapid blood sampling in small animals. *Proc. Soc. Exp. Biol. Med.* **143**, 364.

Schermer, G. (1967). "The Blood Morphology of Laboratory Animals", 3rd edn. F. A. Davis, Philadelphia.

Schmidt, W. K., Tam, S. W., Schotzberger, G. S., Smith, D. H., Clark, R. and Vernier, V. G. (1985). Nalbuphine. *Drug Alcohol Depend.* **14**, 339.

Schneider, D. R., Felt, B. T., Rappaport, M. S. and Goldman, H. (1982). Development and use of a non-restraining waveguide chamber for rapid microwave radiation killing of the mouse and neonate rat. *J. Pharmacol. Meth.* **8**, 265.

Schurig, J. E., Cavanagh, R. L. and Buyniski, J. P. (1978). Effect of butorphanol and morphine on pulmonary mechanics, arterial blood pressure and venous plasma histamine in the anaesthesized dog. *Arch. Int. Pharmacodyn.* **233**, 296.

Schurr, P. E. (1969). Composition and preparation of intravenous fat emulsions. *Cancer Res.* **29**, 258.

Short, C. E. (1987). Pain, analgesics and related medications. *In* "Principles and Practice of Veterinary Anaesthesia" (Short, C. E., ed.), p. 26, 28–46. Williams and Wilkins, Baltimore.

Silverman, J., Huhndorf, M., Balk, M. and Slater, G. (1983). Evaluation of a combination of tiletamine and zolazepam as an anesthetic for laboratory rodents. *Lab. Anim. Sci.* **33**(5), 457.

Smiler, K. L., Stein, S., Hrapkiewicz, K. L. and Hiben, J. R. (1990). Tissue response to intramuscular and intraperitoneal injections of ketamine and xylazine in rats. *Lab. Anim. Sci.* **40**, 60.

Smith, G. (1984). Post-operative pain. *In* "Quality of Care in Anaesthetic Practice" (Lunn, J. N., ed.), Macmillan, London, pp. 164–192.

Steffens, A. B. (1969). A method for frequent sampling of blood and continuous infusion of fluids in the rat without disturbing the animal. *Physiol. Behav.* **4**, 833.

Stillman, R. M., Marino, C. A. and Seligman, S. J. (1984). Skin staples in potentially contaminated wounds. *Arch. Surg.* **119**, 821.

Swaim, S. F. (1980). "Surgery of Traumatized Skin: Management and Reconstruction in the Dog and Cat." W. B. Saunders, Philadelphia.

Tarin, D. and Sturdee, A. (1972). Surgical anaesthesia of mice: evaluation of tribromoethanol, ether, halothane and methoxyflurane and development of a reliable technique. *Lab. Anim.* **6**, 79.

"The UFAW Handbook on the Care and Management of Laboratory Animals" (1987) 6th edn (Poole, T., ed.), Longman Scientific and Technical, Harlow, UK.

Thet, L. A. (1983). A simple method of intubating rats under direct vision. *Lab. Anim. Sci.* **33**, 368.

Tilney, N. L. (1971). Patterns of lymphatic drainage in the adult laboratory rat. *J. Anat.* **109**, 369.

Tomlinson, P. W., Jeffrey, D. J. and Filer, C. W. (1981). A novel technique for assessment of biliary secretion and enterohepatic circulation in the unrestrained conscious rat. *Xenobiotica* **11**, 863.

Trim, C. M. (1983). Cardiopulmonary effects of butorphanol tartrate in dogs. *Am. J. Vet. Res.* **44**, 329.

Trooskin, S. Z., Harvey, R. A. and Greco, R. S. (1983). Prevention of catheter sepsis by antibiotic bonding. *Surg. For.* **34**, 132.

Van Der Meer, C., Versluys-Broers, J. A. M., Tuynman, H. A. R. E. and Buur, V. A. J. (1975). The effect of ethylurethane on haematocrit, blood pressure and plasma glucose. *Arch. Int. Pharmacodyn.* **217**, 257.

Vinegar, R., Truax, J. F. and Selph, J. L. (1976). Quantitative comparison of the analgesic and anti-inflammatory activities of Aspirin, Phenacetin and Acetaminophen in rodents. *Eur. J. Pharmacol.* **37**, 23.

Vinegar, R., Truax, J. F., Selph, J. L. and Johnston, P. R. (1989). Pharmacological characterization of the algesic response to the subplantar injection of serotonin in the rat. *Eur. J. Pharmacol.* **164**, 497.

Waynforth, H. B. (1969). Animal operative techniques (in the mouse, rat, guinea-pig and rabbit). *In* "Techniques in Protein Biosynthesis" (Campbell, P. N. and Sargent, J. R., eds), Vol. 2. Academic Press, New York.

Waynforth, H. B., Hoslman, J. W. and Parkin, R. (1977). A simple device to facilitate intragastric infusion per os in the conscious rat. *Lab. Anim.* **11**, 129.

Wiersma, J. and Kastelijn, J. (1985). A chronic technique for high frequency blood sampling/transfusion in the freely behaving rat which does not affect prolactin and corticosterone secretion. *J. Endocrinol.* **107**, 285.

Wixson, S. K., White, W. J., Hughes Jr, H. C., Marshall, W. K. and Lang, C. M. (1987). The effects of pentobarbital, fentanyl-droperidol, ketamine-xylazine and ketamine-diazepam on noxious stimulus perception in adult male rats. *Lab. Anim. Sci.* **37**(**6**), 731.

Wood, M. and Wood, A. J. J. (1984). Contrasting effects of halothane, isoflurane, and enflurane on in vivo drug metabolism in the rat. *Anesth. Analg.* **63**, 709.

World Health Organization (1978). Inhalation exposure. *In* "Principles and Methods for Evaluating the Toxicity of Chemicals", Part 1, Environmental Health Criteria 6. WHO, Geneva.

Yalkowsky, S. H. (ed.) (1981). "Techniques of Solubilization of Drugs." Marcel Dekker, New York.

EQUIPMENT INDEX

Adjuvants: Freunds; Difco Laboratories Ltd, PO Box 14B, Central Ave, East Molesey, Surrey KT8 0SE, UK
  RIBI; Universal Biologicals Ltd, 30 Merton Road, London SW18 1QY, UK
  Montanide; ISA Seppic, Fairfield, New Jersey, USA
  Titermax; CytRx Corporation, 150 Technology Parkway, Technology Park, Atlanta, Norcross, GA 30092, USA
Anaesthetic induction chamber: Alfred Cox (Surgical) Ltd, Edward Road, Coulsdon, Surrey CR3 2XA, UK
Asep: Galen Ltd, Craigavon, Northern Ireland
Bedding: "Vetbed", Alfred Cox (Surgical) Ltd, Edward Road, Coulsdon, Surrey CR3 2XA, UK
Blood pressure recorder: Harvard Apparatus Ltd, Fircroft Way, Edenbridge, Kent TN8 6HE, UK; Harvard Apparatus Inc., 22, Pleasant Street, South Natick, MA 01760, USA
Bollman cage: International Market Supply, Dane Mill, Broadhurst Lane, Congleton, Cheshire CW12 1LA, UK
Bollman cage: Harvard Apparatus Ltd, Fircroft Way, Edenbridge, Kent TN8 6HE, UK; Harvard Apparatus Inc., 22, Pleasant Street, South Natick, MA 01760, USA
Capsules: Elanco Qualicaps, Lilly Industries Ltd, Kingsclere Road, Basingstoke, Hampshire RG21 2XA, UK
Catheter introducer: Becton-Dickenson UK Ltd, Between Towns Road, Cowley, Oxford OX4 3LY, UK
Coudé (ureteral) catheters: C.R. Bard International Ltd, Forest House, Brighton Road, Crawley, West Sussex RH11 9BP, UK
DecapiCone: Braintree Scientific Inc., PO Box 361, Braintree, MA 02184, USA; Sandown Scientific, 11 Copsem Drive, Esher, Surrey KT10 9HD, UK
Dental Cement: Baxter Dental, Elmgrove Road, Harrow, Middlesex HA1 2HG, UK
Dental drill and burrs: Cottrell and Company, 15 Charlotte Street, London W1, UK

IMP respiratory monitor: Veterinary Instrumentation, 50 Broomgrove Road, Sheffield S10 2NA, UK

Incubators: Harvard Apparatus Ltd, Fircroft Way, Edenbridge, Kent TN8 6HE, UK; Harvard Apparatus Inc., 22, Pleasant Street, South Natick, MA 01760, USA; Department of Medical Physics, Royal Postgraduate Medical School, Ducane Road, London W12 0HS, UK

Infra-red lamp: International Market Supply, Dane Mill, Broadhurst Lane, Congleton, Cheshire CW12 1LA, UK

Inhalation dosing: Braintree Scientific Inc., PO Box 361, Braintree, MA 02184, USA

KY jelly: Johnson & Johnson Ltd, Maidenhead SL6 3UG, UK

L-Cath catheter: International Market Supply, Dane Mill, Broadhurst Lane, Congleton, Cheshire CW12 1LA, UK

Lymphoprep: Vestric Ltd, Lockfield Avenue, Enfield, Middlesex EN3 7QR, UK

Melolin: Smith and Nephew Pharmaceuticals Ltd, Bessemer Road, Welwyn Garden City, Hertfordshire, UK

Metabolism cages: North Kent Plastic Cages Ltd, 1 Bilton Road, Erith, Kent DA8 2AN, UK; Labcare Precision Ltd, Calleywell Barn, Coldwell Lane, Aldington, Ashford, Kent TN25 7DX, UK; Harvard Apparatus Ltd, Fircroft Way, Edenbridge, Kent TN8 6HE, UK; Harvard Apparatus Inc., 22, Pleasant Street, South Natick, MA 01760, USA; Jencons Scientific, Cherry Court Way Industrial Estate, Stanbridge Road, Leighton Buzzard, Bedfordshire LU7 8UA, UK

Microchip identification system: Animalcare Ltd, Common Road, Dunnington, York, UK; Biomedic Data Systems Inc., 255 West Spring Valley Avenue, Maywood, NJ 07607, USA

Microvette: Sarstedt Ltd, 68 Boston Road, Beaumont Leys, Leicester LE4 1AW, UK

Monoject (insulin syringe): Sherwood Medical Industries Ltd, London Road, County Oak, Crawley, West Sussex, UK

Mouth gag: International Market Supply, Dane Mill, Broadhurst Lane, Congleton, Cheshire CW12 1LA, UK

Nebulisers: Devilbiss Healthcare: Unit 4, Aerial Way, Great Southwest Road, Hounslow, Middlesex TW4 7JW, UK; Schuco International London Ltd, Lyndhurst Avenue, Woodhouse Road, London N12 0NE, UK

Nylon bolting cloth: Thomas Locker, PO Box 161, Church Street, Warrington, Cheshire WA1 2SU, UK

Operating boards/tables: Harvard Apparatus Ltd, Fircroft Way, Edenbridge, Kent TN8 6HE, UK; Harvard Apparatus Inc., 22, Pleasant Street, South Natick, MA 01760, USA

Opsite: Smith and Nephew Pharmaceuticals Ltd, Bessemer Road, Welwyn Garden City, Hertfordshire, UK

Oral (gavage) cannula: International Market Supply, Dane Mill, Broadhurst Lane, Congleton, Cheshire CW12 1LA, UK; Harvard Apparatus Ltd,

LE9 6PD, UK

Tissue adhesive (Vetbond): Alfred Cox (Surgical) Ltd, Edward Road, Coulsdon, Surrey CR3 2XA, UK; Animal Care Products/3M, St Paul, MN 55144, USA

Trocar: International Market Supply, Dane Mill, Broadhurst Lane, Congleton, Cheshire CW12 1LA, UK

Trypan blue (and other dyes): Raymon A. Lamb, 6 Sunbeam Road, London NW10 6JL, UK

Tubing (catheters): Portex Ltd, Hythe, Kent CT21 6JL, UK

Vacutainer: Becton-Dickenson UK Ltd, Between Towns Road, Cowley, Oxford OX4 3LY, UK; (Vacuette); Greiner Labortechnik Ltd, Station Road, Cam, Dursley, Gloucestershire GL11 5NS, UK; (Monovette); Sarstedt Ltd, 68 Boston Road, Beaumont Leys, Leicester LE4 1AW, UK

Vascular access port: Norfolk Medical Products Inc., 7307 N Ridgeway, Skokie, IL 60076, USA; Braintree Scientific Inc., PO Box 361, Braintree, MA 02184, USA; Sandown Scientific, 11 Copsem Dr., Esher, Surrey KT10 9HD, UK

Ventilators: Harvard Apparatus Ltd, Fircroft Way, Edenbridge, Kent TN8 6HE, UK; Harvard Apparatus Inc., 22, Pleasant Street, South Natick, MA 01760, USA

"Vetbed": Alfred Cox (Surgical) Ltd, Edward Road, Coulsdon, Surrey CR3 2XA, UK

Wire mesh sieves: Endecotts Ltd, Lombard Road, London SW19 3BR, UK

Anaesthetic Drugs—UK and USA generic names, trade names and manufacturers

UK approved name	UK trade name	Manufacturer	USA approved name	USA trade name	Manufacturer
Acepromazine maleate	Acetylpromazine	C-Vet	Acepromazine maleate	Acepromazine	TechAmerica Group Inc.
Alpha-chloralose	—	BDH Ltd	Alpha-chloralose	—	Fisher Scientific Co.
Alphaxalone/ alphadolone	Saffan	Glaxovet	Alphaxalone/alphadolone	—	—
Alfentanil	Rapifen	Janssen	Alfentanil	—	—
Atropine sulphate	—	Antigen Ltd	Atropine sulphate	—	Fort Dodge Labs
Azaperone	Stresnil	Janssen	Azaperone	Stresnil	Pitman-Moore
Bupivicaine hydrochloride	Marcain	Astra	Bupivicaine hydrochloride	Marcaine Sensorcaine	Winthrop Labs
Buprenorphine	Temgesic	Reckitt and Colman	Buprenorphine	—	—
Butorphanol	Torbugesic Torbutol	C-Vet	Butorphanol	Torbugesic Torbutol	Bristol Labs
Diazepam	DF-118	Duncan, Flockhart & Co. Ltd	Diazepam	Valium	Roche

Cont'd.

Anaesthetic Drugs. Continued

UK approved name	UK trade name	Manufacturer	USA approved name	USA trade name	Manufacturer
Doxapram hydrochloride	Dopram-V	Willows	Doxapram hydrochloride	—	A. H. Robbins Co.
Enflurane	Ethrane	Abbott	Enflurane	Enflurane	Abbott
Ether	—	May & Baker Ltd	Ether	—	Squibb Pharmaceutical Co.
Etomidate	Hypnomidate	Janssen	Etomidate	Amidate	Abbott
Etorphine-methotrimeprazine	Immobilon S.A.	C-Vet	—	—	—
Fentanyl	Sublimaze	Janssen	Fentanyl	—	—
Fentanyl-droperidol	Thalamonal	Janssen	Fentanyl-droperidol	Innovar Vet	Pitman-Moore Co.
Fentanyl-fluanisone	Hypnorm	Janssen	—	—	—
Flumazenil	Anexate	Roche	Flumazenil	—	—
Halothane	Fluothane	ICI, May & Baker Ltd	Halothane	Fluothane	Fort Dodge Labs
Isoflurane	Forane	Abbott	Isoflurane	Forane	Ohio Medical Products

Ketamine	Vetalar, Ketalar	Parke-Davis, Parke-Davis	Ketamine	Vetalar, Ketaset	Parke-Davis, Bristol Labs
Lignocaine	Xylocaine	Astra, UK	Lidocaine	Xylocaine	Astra Pharmaceutical Products, USA
Medetomidine	Domitor	Smith-Kline	Medetomidine	—	—
Methohcxitone	Brietal	Lilly	Methohexital	Brevital	Lilly
Methoxyflurane	Metofane	C-Vet	Methoxyflurane	Penthrane, Metofane	Abbott Labs, Pitman-Moore
Midazolam	Hypnovel	Roche	Midazolam	Versed	Hoffman-La Roche
Naloxone	Narcan	Du Pont	Naloxone	Narcan	Endo Laboratories Ltd
Pentazocine	Fortral	Sterling Research (Winthrop)	Pentazocine	Talwin-V	Winthrop Labs
Pentobarbitone	Sagatal	May & Baker Ltd	Pentobarbital	Nembutal	Abbott Labs
Pethidine	—	—	Meperidine	Demerol	Elins-Sinn Inc
Propofol	Diprivan	ICI	Propofol	—	—
Thiopentone	Intraval	May & Baker Ltd	Thiopental	Pentothal	Abbott Labs
Urethane	—	—	Urethane	—	Merck & Co.
Xylazine	Rompun	Bayer	Rompun	—	Haver

SUBJECT INDEX